manual as well as a guide for emerging practitioners seeking to transition into a field that is yet to be defined and that offers the promise of a new era of human habitat making as a direct response to the looming ecological crisis.

Assia Crawford is Assistant Professor in the College of Architecture and Planning at the University of Colorado Denver, USA. Her creative practice research focuses on the development of biological material alternatives and digital fabrication practices for a post-Anthropocene era. Her work sits on the intersection of architecture, science and critical theory, and employs experimental and speculative design to address ecological challenges faced by communities at a time of environmental uncertainty.

Assia is a registered architect with Architects Registration Board (ARB) the UK's regulator of architects and holds a PhD in architecture. She has previously held positions as the architect for the Hub for Biotechnology in the Built Environment (HBBE) and artist-in-residence at the Wellcome Centre for Mitochondrial Research. Assia is an editor of *Biotechnology Design Journal* and runs Wild Futures Lab, a research and teaching bio-design fabrication wet lab, which explores fabrication and making in the age of environmental decline.

Bio Design

Series editor: Martyn Dade-Robertson,
Newcastle University, UK

The Bio Design series offers the opportunity for designers from fields as diverse as architecture, fashion design and product design to present and explore designs and design research, which use living systems as part of their production and operation. The series offers readers in depth project descriptions, analysis of processes and the intellectual contexts of Bio Design. Such explorations have not, until now, been made available in long form. The series also allows designers to explore the potentials and challenges of bio-design as an emerging field of design and research.

The Bio Design book series will distinguish between key areas within the field, including design fictions, biomimicry, bioinspired design and the use of biological materials and systems. While open to a range of voices, the series will, as a collection, offer unified framework for thinking about Bio Design opening up this new rapidly growing field to a new generation of designers and researchers.

Living Construction
Martyn Dade-Robertson

Designer's Guide to Lab Practice
Assia Crawford

Designer's Guide to Lab Practice

Assia Crawford

LONDON AND NEW YORK

Designed cover image: "Mitochondrial Matrix" © Assia Crawford

First published 2024
by Routledge
4 Park Square, Milton Park, Abingdon, Oxon OX14 4RN

and by Routledge
605 Third Avenue, New York, NY 10158

Routledge is an imprint of the Taylor & Francis Group, an informa business

© 2024 Assia Crawford

The right of Assia Crawford to be identified as author of this work has been asserted in accordance with sections 77 and 78 of the Copyright, Designs and Patents Act 1988.

All rights reserved. No part of this book may be reprinted or reproduced or utilised in any form or by any electronic, mechanical, or other means, now known or hereafter invented, including photocopying and recording, or in any information storage or retrieval system, without permission in writing from the publishers.

Trademark notice: Product or corporate names may be trademarks or registered trademarks, and are used only for identification and explanation without intent to infringe.

British Library Cataloguing-in-Publication Data
A catalogue record for this book is available from the British Library

ISBN: 9781032426860 (hbk)
ISBN: 9781032426846 (pbk)
ISBN: 9781003363774 (ebk)

DOI: 10.4324/9781003363774

Typeset in Calvert
by codeMantra

To my family

Contents

Series Editor's Preface		xii
Ethics and Society		xiv
Acknowledgments		xvi
1	**This Is Not the Beginning**	2
	Growing Materials	4
	Cell Culture	5
	Fungus	6
	Bacteria	9
	Algae	11
	Building Fabric	13
2	**Design in the Lab**	22
	Approaches in Bio-Design	23
	Home Labs and Experimental Design Spaces	23
	Working in a Science Lab: Shadowing, Collaboration and Conversation	25
	Health, Safety and Waste Disposal	27
	Setting Up a Design Wet Lab	28
	Contamination and Aseptic Conditions	31
	Species Matter	36
	Designing a Process	37
	Working in Multiples, Experiments and Setting Up Controls	38
	Viewing the World	39
	Imaging Larger Samples Areas	39
	Imaging Individual Cells	41

Surface Imaging		42
Spectroscopy		42
Imaging Photosynthetic Organisms		43
3	**Working Methods**	48
Working with Fungus		49
Fungal Species: Sourcing and Considerations		49
Liquid Culture		51
Agar Petri Dishes and Slants		53
Grain Spawn Preparation		55
Bulk Substrate Preparation		55
Working with Microalgae and Cyanobacteria		57
Cultivation Modes		57
Microalgae Suspension Cultivation Systems		58
Sourcing Microalgae Culture		59
Cultivation and Incubation		59
Media Preparation		59
Nutrient Sources		61
Microalgae Harvesting		64
Working with Bacterial Cellulose		65
Strains		66
pH		67
Static vs. Agitated Culture and Oxygenation		67
Temperature		68
Fermentation Time		68
Nutrient Media		69
Harvesting, Drying and Posttreatments		70
4	**Experiments in Design**	80
Mycelium Casting		81
Student Project: Mycelium Making		85
Extrusion-Based 3D Printing		88

Clay 3D Printing for Coral Restoration	88
Bio-composites and Photosynthetic Organisms	97
3D Printing with Bio-laden Hydrogels: Photosynthetic Textile Bio-composites	98
3D Ceramic Photosynthetic Bio-composites	105

5 Where to Next? — 118

The Consciousness Shift — 121

Plurality of Practice — 123

Constructing Networks: the Relationship between Scientific and Creative Practice — 124

Glossary — 128

Index — 132

Series Editor's Preface

The emergence of biotechnologies and their integration into human-centered contexts of use has led to a new design paradigm known as "Bio Design". Bio Design, bio-design or biodesign, depending on the source, has a wide range of definitions. Early examples of Bio Design are mainly focused on the design of medical devices and methods to support tissue engineering. Bio Design has also been used alongside fields such as synthetic biology. This series, however, uses a broader definition of Bio Design which includes the integration of biological processes across a range of creative design fields including, but not limited to, architecture, fashion and apparel design, product design and interaction design. Too often, real applications and technologies are confused with grandiose pronouncements of technological revolution based on science fictions built on limited or no real engagement with messy biological reality. The series will allow for deeper explorations of Bio Design projects and theories. The books will necessitate critical and reflective descriptions of the Bio Design field focused on the underlying processes, methods and theories. In defining the territory of this book and the series it initiates, we are seeking a happy middle ground between design speculation and grounded experiment, between critical thinking and creative naivety, between formal elegance and radical complexity, and to provide ways of thinking which will lead to critical ways of making.

Martyn Dade-Robertson
Newcastle University
February 2020

Ethics and Society

Any conversation on designing with biology necessitates refection about ethics. Novel biotechnology and bioengineering applications have the potential to provide enormous benefits for society. However, the relationship between human activities and the current climate emergency due to global warming remind us that technologies used to solve one set of problems are capable of creating many others. Beyond environmental impacts, there is also the challenge of considering diverse political and social values in the development of bioengineered technologies and materials. As a result, we wanted to establish an ethical position in developing this series.

- We will not seek to publish material which could be applied directly to the development of weapons or deliberate causes of harm.

- Where experiments or processes are introduced that may be harmful to the individuals conducting them or the environment, we will make these risks clear, especially given audiences who may be unfamiliar with the techniques and technologies being described.

- Authors in the series will be required to confirm (where applicable) that appropriate risk assessment, ethics review, informed consent and animal welfare protocols have been met, in compliance with local institutional and governmental regulations. While we make every effort to anticipate risks in our research, the unintended consequences of technology are harder to predict. However, in formulating this series, we believe that Bio Design is at its best when it is a reflective and inclusive practice where ethical and responsible principles are embedded. The books and editorial guidance in this series prioritize this, while accepting diversity of opinion and position.

Martyn Dade-Robertson
Carmen McLeod
Newcastle University
February 2020

Acknowledgments

I would like to express my gratitude to Dr Gary Caldwell, a PhD supervisor and mentor from the School of Marine Science at Newcastle University, and Pichaya In-Na, my inspirational collaborator and partner in crime. They laid the groundwork for the photosynthetic living materials research highlighted in this book and opened up to me the wondrous world of science practice. Above all, I am grateful for their patience, their welcoming supportive spirit and their friendship, which have made this research a joyful and fulfilling pursuit.

Furthermore, I would like to thank my supervisor Prof. Rachel Armstrong, a pioneer in the bio-design field and a much needed role model for the new generation of interdisciplinary practitioners. I would also like to express my gratitude to Dr Ben Bridgens, my PhD supervisor and mentor for his levelheaded and precision-driven approach to research that helped me bridge the design and science divide. I would also like to thank Dr James Guest and Adriana Humanes for allowing me to collaborate on the CoralPlugs project as well as opening up to me the enigmatic world of corals. I am also deeply grateful to the bio-design community that has emerged from the Hub for Biotechnology in the Built Environment, where I found a community of likeminded misfits as well as a new sense of belonging.

I would also like to thank my students for their participation in the Mycelium Making design studio and their joy of discovering lab practice, which shaped my intention for this book to be a guide for this new generation of thinkers and makers as they transition on a path less traveled.

I would also like to express my gratitude to Prof. Martyn Dade-Robertson, the editor of this much needed series of books, and Fran Ford, an editor at Routledge for trusting me with my first book.

This Is Not the Beginning

Chapter 1

DOI: 10.4324/9781003363774-1

As architects and designers, we often see ourselves as creators, and we claim ownership of products, processes and ideas. This claim to authorship has at times been extended to living organisms in the form of genetically modified ornamental fish (Birke, 2009), glow in the dark bunnies (Best, 2009), patented bacteria (*Diamond v. Chakrabarty*, 1980) and designer pathogens (Roosth, 2017). These Doctor Moreau type worlds are often sensationalized in the media and passed off as the panacea to problems facing humanity. If we could swap out genes the way we edit code, redacting what we deem to be redundant and plugging in blocks of information, then we would have finally established our dominion over nature as envisioned by Francis Bacon over 400 years ago in his chillingly prophetic work The New Atlantis (Bacon, 1626).

This is often the lens through which designers emerging from a traditional design background view bio-design prior to embarking on working with living organisms. I, having gone through that journey and remember early conversations with biologists, inorganic chemists, chemical engineers and marine scientists to that effect. I recall sitting in offices interrogating scientists about their work, having read a handful of academic papers on gene editing, and modified microbial organisms that were promising to be the future of remediating toxins, producing biofuels and even making other planets habitable. These early encounters with level-headed individuals, who had grappled with the reality of such work for decades, left me questioning, what I could contribute as an architect to problems I had reductively assumed I could solve by rearranging parts, the way I would move lines and elements within my drafting software.

I grew weary of computational and mechanistic metaphors of building blocks of living organisms. It was not only the oversimplification of processes that are in fact incredibly complex and interconnected but also a disconnect from the reality of another living organism that assumes the role of inanimate matter that became problematic in my mind. It was not until I discovered books such as *Staying with the Trouble* (Haraway, 2016) and *The Death of Nature* (Merchant, 1980) that I began to make sense of how other modes of thinking about life could shape design practice. It became apparent that working with living organisms is an act of co-creation with actors that must retain their agency, if we are to begin to address the causes of the anthropogenic problems we are trying to solve. This line of thinking also highlights the agency of the designer as a mediator between the sciences and the ethical implications discussed within the humanities and social sciences. In the words of Donna Haraway,

> It matters which stories tell stories, which concepts think concepts. Mathematically, visually, and narratively, it matters which figures figure figures, which systems systematize systems.

(Haraway, 2016, p. 101)

As a designer, I realized, I have a different role to play. My work, although employing scientific methods, is not pure science, and it would never be interchangeable for several reasons. First, although I may mimic the motions and follow protocols within a lab setting and obtain data in a similar way, I do lack foundational understanding, the in-depth expertise and the skills. The second reason rests in the fact that a design practitioner comes into such settings with a different set of objectives and different ways of connecting tools and thinking from outside the sciences, with a habit of subverting methods, placing them outside traditional constraints. For example, in Chapter 4, I talk about this hybridization of methods to achieve design goals and to accelerate testing along a trajectory of design solutions. The experiments described often skip steps within the scientific process and leave potential scientific questions unanswered. However, there is a shift between scales that takes place, in particular moving from a petri dish to architectural bio-fabrication. Another important focus is the drive toward an *in vivo* functionality. While most fundamental studies occur within controlled environments with a small number of variables influencing the experiment, there is an underlying push to move outside the bounds of the *in vitro* setup. This also differentiates some of the tools and methods disused in Chapter 3 that borrow from other disciplines such as mushroom farming and food fermentation.

GROWING MATERIALS

Emerging practices and research that view the environment as a network are challenging the mechanical view of the world and exclusionary approaches to hygiene proposed by the Modern Movement (Tostões, 2020). Autopoiesis, the ability to self-generate and self-repair, are a captivating model that is displacing the notion of static, inanimate matter in design. Autopoiesis being a core biological principle of a self-sustaining organism is also accompanied by symbiosis, the notion of mutualistic partnerships between species. This is key in understanding the living organisms that we employ within design. For a long time, the notion of antibiotic spaces was deemed to be the model of health. Similarly, cultivating organisms in isolation has been the primary mode of fundamental laboratory studies where species are grown independent of other living organisms. However, the natural complexity of uncontrolled environments can be both a challenge and an opportunity for living materials.

If we are to look at living materials, we are faced with the by-products of a living metabolism. These may be bacterial cellulose, calcite, oxygen or the dead or dormant matter of the microorganism, in the form of biomass. In certain instances, we focus on guiding the development of the organism until it reaches a desired state, at which point it is immobilized halting its development, as in the case of growing a mycelium product. On the other hand, there is an emphasis on sustaining the organism to benefit from its metabolic functions, for example, photosynthetic façades that sequester carbon dioxide (CO_2) and produce oxygen. These fundamental differences in cell mass vs. metabolic function influence both the methods and the thinking surrounding such work.

Using microorganisms for their structural makeup and metabolic activities, for instance, is not a unique notion. In health care, construction, nanotechnology, and material manufacturing, scientists have attempted to maintain organisms and influence their behavior by manipulating their environment. There are different types of practice and application based on various organismal types and processes. The following sections offer an overview of various generative examples.

Cell Culture
The principle of cell culture is at the heart of most bio-design processes. It involves the developing or identifying appropriate mediums and the correct incubation environment conditions for growing single cell organisms or living cells of multicellular organisms of plant or animal origin.

In 1912, an experiment took place within the Rockefeller Institute for Medical Research, where an incubated egg was opened prior to hatching and the host's heart was removed. The heart was then grown artificially by supplying it with the necessary nutrients for growth. The artificially cultivated organ exhibited superior growth compared to a naturally developed one. Hence, there have been numerous experiments and speculations as to the idea of artificially growing animal and plant cells, both within medicine (Ngan *et al.*, 2019) and the food industry (Dick, Bhandari and Prakash, 2019), even construction (Tandon and Joachim, 2014).

Tissue culture cultivation of animal cell products including meat or leather often utilizes stem cells and animal-derived nutrients. *Victimless Leather*, an art installation by Oron Catts and Ionat Zurr, explores the moral repercussions of fabricating with animal tissue culture. The work sheds light on the use of animal-derived product within the process and acknowledges the deeper implications of this mode of fabrication. The cultivation process perpetuates growth until the supply of nutrients is terminated, halting the growth and in effect "killing" the tissue (Catts and Zurr, 2004).

Cell culture techniques can also be applied to plant cells to grow plant-based alternatives such as artificial wood without the need for cutting down mature trees. Researchers at Massachusetts Institute of Technology (MIT) successfully cultured *Zinnia elegans* cell, present in a flower, and have since studied the potential for growing wood-like materials with tenable mechanical properties (Beckwith, Borenstein and Velásquez-García, 2022). The culture that the cells were grown within is a sugar-based solution pointing to a demand for either a waste stream compatible with this particular species or the need for agricultural produce that places pressures on the environment. However, the benefit of such timber alternatives is their ability to be molded into desired products via 3D printing or casting without the restrictions of conventional timber and without material wastage.

Fungus
The fungal kingdom merits its own branch on the evolutionary tree and has permeated most landscapes, be it natural or social. There are an estimated 1.5 million fungal species on the planet, yet a mere 80,000–120,000 have been recorded (Hawksworth, 2001; Kirk *et al.*, 2001). These living organism has survived mass

extinctions, transformed the rocky makeup of the Earth into fertile ground, and decomposed waste, freeing up resources for new life to emerge (Sheldrake, 2020). Myco-design, a branch of bio-design that can be defined as a separate creative subfield, is growing rapidly due to the sustainability aspects of growing materials using fungus. The cultural and ecological potentials of fungi have become the central theme of various studies conducted by scientists and anthropologists captured in books such as *The Mushroom at the End of the World* (Tsing, 2017), *Entangled Life* (Sheldrake, 2020), *In Search of Mycotopia* (Bierend, 2021) and *The Origins of Sociable Life* (Hird, 2009). These books have made the world of fungi and their role as bio-remediators, decomposers, and social catalysts widely accessible to a non-specialist audience.

Mycelium, the vegetative part of the fungus, made up of branching structures called hyphae, can be grown on a variety of substrates as it can digest a range of organic compounds such as coffee grounds, saw dust, and cardboard, to name a few. The growth rate of the mycelium and the material properties of the final product are determined by the type of substrate and mushroom species that the mycelium spore has been harvested from, as well as temperature, light levels and humidity (Stefanova, *et al.,* 2019). Artists and commercial companies have manipulated these parameters to produce a wide array of products such as mycelium "leather" by MicoWorks (Ross, 2019) or myco-bricks that have compressive strength like that of timber (Xing *et al.,* 2018). These novel materials are a continuation of a longstanding practice of utilizing microbial metabolic functions for human purposes such as food and beverage fermentation.

Mycelium has presented designers with the possibility of making compostable building blocks and products on a human scale. An example of such applications is the work of designer Eric Klarenbeek, who utilizes digital fabrication of living mycelium printed alongside a biodegradable polymer shell from a double extrusion 3D printer (Fairs, 2013). The printed product is further incubated until the mycelium encapsulates the polymer. The mycelium continues to grow until it is dried out and fired in an oven; in essence, killing off the living organism once the product is fully "grown". The works are biodegradable and have a carbon-negative footprint. This type of additive manufacturing opens new possibilities of fabricating with a wide range of living organisms and gives microorganisms agency within the fabrication process.

Other research that captures work with fungi in design includes the work of Asya Ilgun (Ilgun and Schmickl, 2022) who explores the relationship between fungi and bee health, Dilan Ozkan (Ozkan *et al.*, 2022) who explores the parametric nature of mushrooms and Andrew Adamatzky (Adamatzky and Gandia, 2022; Adamatzky, Goles, *et al.*, 2022) and Pill Ayers (Adamatzky, Ayres, *et al.*, 2022) who study the computational potentials of mycelium. Such studies challenge our established perceptions of design as a human-centered outcome-driven pursuit. Other practitioners focus on the structural and material properties of mycelium products as a sustainable alternative to traditional building materials. Examples include Elise Elsacker and Adrien Rigobello who investigate material properties of myco-materials (Elsacker *et al.*, 2019; Elsacker, Søndergaard, *et al.*, 2021; Elsacker, Vandelook, *et al.*, 2021; Rigobello, Gaudillière-Jami and Ayres, 2022), Felix Heisel (Heisel *et al.*, 2017) and Johnathan Dessi-Olive (Dessi-Olive and Hsu, 2021) who develops structural prototypes with mycelium composites. There are also speculative works that invite reflection into the potentials of scaling of myco-solutions, such as the work of Claudia Colmo who investigates the bio-remediating potentials of mycelium in landscape design (Colmo and Ayres, 2022) and Svenja Keune who speculates as to the potentials of co-creating with other species (Keune, 2021).

There are various sources for bio-designers entering the field that offer an overview into bio-design such as *Bio Design: Nature + Science + Creativity* (Myers and Antonelli, 2014), *Biomimicry in Architecture* (Pawlyn, 2016), *The Neri Oxman: Material Ecology Catalogue* (Antonelli, 2020), *Materials Experience: Fundamentals of Materials and Design* (Karana, Pedgley and Rognoli, 2014) and *Materials Experience 2: Expanding Territories of Materials and Design* (Pedgley, Rognoli and Karana, 2021). However, when researching working methods, in addition to emerging bio-design research, designers often venture into other related fields to source and improve working methods in their own practice. When it comes to fungi, some of the most accessible examples for bio-designers come from the world of mushroom growing and mushroom farming. Such sources offer a robust introduction to the basic principles of cultivating fungi and understanding their behavior in nature. Examples include *Radical Mycology* (McCoy, 2016), *Mycelium*

Running (Stamets, 2005) and *The Mushroom Cultivator* (Stamets and Chilton, 1983). As designers become more familiar with scientific principles and scientific writing, publications from the world of mycology become indispensable sources of information that offer experimental method and analysis techniques that allow practitioners to develop their own experiments. Examples of such works include *Fungi: Experimental Methods in Biology* (Maheshwari, 2016), *Introduction to Mycology* (Gnanam, 2019) and *Fungi: Biology and Applications* (Kavanagh, 2017).

Bacteria

Bacteria are often viewed as the sculptors of the planet (Hird, 2009), although traditionally stereotyped as pathogenic (Lorimer, 2020). Our struggle to conquer bacteria and remove them from our immediate surroundings can be traced back to early humans and the struggle to preserve food. Various antibiotic treatments emerged to combat pathogenic agents, including using heat to cook, applying alcohol and honey and changing the pH by adding salt to preserve food.

However, as we transitioned into an agrarian society, fermentation became a common practice to produce cheese, yogurt, mead and other types of fermented beverages (Amyes, 2013). These examples illustrate the early social evolution of our relationship with bacteria; however, a more poignant entanglement with unicellular life is our own physiological makeup. Symbiotic relationships with microorganisms are at the very core of our own human existence and the health of the planet. Large animals, including humans, are not independent entities but holobionts, made up of and supported by symbiotic relationships (Crawford, 2022). Vital functions within the body depend on bacterial communities, with approximately 10,000 microbial species forming the human microbiome (The Human Microbiome Project, 2012). Prior to the Cambrian Period, life on the planet was exclusively unicellular and responsible for the evolution of metabolic functions vital to all organisms today (Furusawa and Kaneko, 2002). Bacteria believed to have originated some 3.5 billion years ago (Brown and Doolittle, 1997) are responsible for shaping a hospitable setting for multicellular life such as humans and other complex life-forms. About 1.8 billion years ago, cyanobacteria drastically altered the atmospheric composition of the planet, by evolving to produce oxygen (Battistuzzi, Feijao and Hedges, 2004). This is just one example of a long line of functions that evolved during that period such as sensory responses, light energy conversion, all reproductive modes, community organization, communication and chemical conversion (Hird, 2009). These interspecies relationships are the subject of investigation of various books, namely *I Contain Multitudes* (Yong, 2016), *The Probiotic Planet* (Lorimer, 2020) and *The Origins of Sociable Life* (Hird, 2009).

Advances in scientific and ecological thinking and the invention of more precise instruments for viewing the world such as the microscope are reshaping our definition of microbial life (Lowenfels and Lewis, 2006; Stefanova, 2021b). By harnessing metabolic functions or their by-products, we can substitute mechanical function with biological alternatives. Bacteria are continually adapting and responding to their environment, which, on a microscale, presents rapidly changing treacherous landscapes. This organismal resilience and responsive behaviour present opportunities to choreograph processes such as the production of sensors, bio-mineralization of sand to produce masonry structures, healing concrete (Fisher, 2021), 3D printing of responsive materials, bovine leather alternative in the form of bacterial cellulose, etc. (Chebath-Taub *et al.*, 2012; Schaffner *et al.*, 2017; Arnardottir *et al.*, 2020). Bacterial metabolic products present opportunities to reshape traditional production. Bacterial products, such as calcium carbonate utilized in bacterial mineralization or bacterial cellulose, often explored as a substitute for traditional textiles or alternatives to food production such as bacterial ink (Myers and Antonelli, 2018), captivate the imagination of designers and open the gateways to a new type of human environment. Projects include bacterial ink developed by Ori Elisar (Elisar, 2018), bioluminescent lighting using octopus's bacteria by Teresa van Dongen (van Dongen, 2015), BioCouture designed by Suzanne Lee and microbial fuel cells that utilize bacteria to produce energy (Mohammadifar *et al.*, 2018). Many of these emerging technologies may sound like science fiction and are oftentimes present in fledgling studies that explore the potentials of these materials in a future that challenges inert products. However, there are also attempts to industrialize bacterial products for single-use packaging, medical applications and concrete repair.

The various outcomes and modes of working with bacteria span a wide range of practices and methods. The literature is currently growing to guide scientists and designers in diverse techniques for working with bacteria for the purposes of biomineralization, such as self-healing concrete (Fisher, 2021). Methods for growing bacterial cellulose are captured in *Bacterial Cellulose Sustainable Material for Textiles* (Muthu and Rathinamoorthy, 2021) or working with bioluminescence (Santhanam, 2022).

There is a growing body of speculative, experimental and artistic projects that investigate co-creative fabrication practices. They are often underlined by critical discourse that brings into question genetic modification such as the work of Carole Collet (Collet, 2020), the future of natural and engineered ecologies (Armstrong, 2018) and multispecies entanglements with microbial life (Stefanova, 2021b). The abundance of creative work and diversity of practice is indicative of a generation of designers and scientists who are looking to microbial life to generate solution-oriented partnerships as well as reflective practices.

Algae

Algae are responsible for creating the oxygen-rich atmosphere that has allowed life on this planet to flourish. Their abundance and diversity provide an ample source for design and science exploration, which is already widely evident in prototypes and working examples, that harness algal metabolic processes from their ability to sequester carbon dioxide and other pollutants to their potential to generate electricity and biomass for fuel and food. There are around 50,000 phytoplankton species that have evolved to live in a range of conditions and are found in most places around the world (Reynolds, 2006). Despite their ability to adapt gradually to different environmental conditions, phytoplankton species have developed to live in specific environments. Therefore, any sudden changes to the environment can result in death or underperformance. To add to that, various species are better suited for distinct applications such as CO_2 fixation, wastewater remediation and biomineralization, to name a few.

Waste is generated by most industries, resulting in deterioration of water, air, soil and natural habitats. In the context of the built environment, pollution is an inevitable by-product of the creation of our artificial habitats. Buildings begin to contribute to pollution from their inception and continue to have an impact long after their demolition (Cairns and Jacobs, 2014). It is estimated that construction is responsible for 47% of CO_2 emissions within the UK, with the "in-use" phase contributing to the majority of CO_2 production, accounting for 83% of the total amount, followed by "manufacture", which makes up 15% (BIS, 2010). Algae have a high rate of CO_2 sequestration, on average 1.8 kg of CO_2/1 kg biomass and requiring less space and fewer virgin resources for growth. Their presence in most environments on the planet and bioremediation potential presents a potent ground for bio-design exploration. Several different systems have been developed to address the issues of air pollution, including placing plants in offices, double skin façades, and integration of faster growing or more efficient organisms. An example is the proposed introduction of Azolla, the fastest-growing plant species that can grow as much as 100 g/m² per day during peak conditions (Parhizkar and Afghani Khoraskani, 2017). Such studies point to the need to identify species with superior performance that can perform those functions in flexible conditions, thus eliminating the need for wet environments, regular introduction of nutrients or space-hungry setups.

A popular solution to living algae integration into buildings are photo-bioreactors that utilize phototropic microorganisms such as algae and cyanobacteria to produce useful outputs such as biomass and oxygen. As microalgae require light for photosynthesis, many of the proposed applications look to façade integration, and the liquid culture is typically cultivated in transparent containers that circulate nutrients, air and water. Such systems also filter and harvest excess organic matter on-site. However, this is a mechanically demanding model. This is an approach explored by prominent researchers in the field such as Kyoung Hee Kim, who develops air quality enhancement systems, documented within her book, entitled *Microalgae Building Enclosures* (Kim, 2022).

An example of integration of photo-bioreactor technology into the building fabric is the *BIQ Building* by Splitterwerk Architects and Arup, which features a photo-bioreactor façade design that provides electrical power to a residential building through the production and utilization of biomass. Along with rigid containers that can be limiting in terms of integrating within the building fabric, experimental material systems are beginning to emerge, such as Petra Bogias's Algae Textile, a proposal for a lightweight photo-bioreactor made up of a flexible skin (Bogias, 2014) and the Algaerium by Marin Sawa (Sawa, 2012), which features the use of algae slurry as a substitute to conventional ink in inkjet printing. In a similar vein, prototypes by ecoLogicStudio, such as Vertical Algae Garden, HORTUS and Tree.ONE feature various bioreactor designs within speculative designs (EcoLogicStudio, 2021). In these examples, living algae are cultivated and sustained on-site for the purposes of CO_2 sequestration, bioremediation and biomass production within urban centers. This type of small-scale algae cultivation, compared to industrial algae cultivation comes with inherent challenges of maintenance, specialist equipment, monitoring requirements and limited yields against high costs. Yet these examples are a vital part of the transition into a living building realm, necessitated by steep environmental decline.

In addition to living algae cultivation, there is also a growing market for biological substitutes to petrochemical products. In this context, algae biomass becomes a substitute or supplement for biodegradable and eco-friendly alternatives. Experimental materials include 3D printing materials (Stefanova *et al.*, 2021; Mandal *et al.*, 2023), edible packaging (Senarathna, Wickramasinghe and Navaratne, 2022), biological yarn (Keel Labs, 2022), algae biomineralized masonry products (Heveran *et al.*, 2020) and algae-based paint (Bouillon, 2012). This is an increasingly growing palette of alternatives that look to phytoplankton for carbon negative solutions.

BUILDING FABRIC

British architect Frank Duffy proposed that the building could be defined as "several layers of longevity of building components" (Braham, Hale and Sadar, 2007, p. 334). For Duffy, "Time is the essence of the real design problem" (Braham, Hale and Sadar, 2007, p. 335). He divided these layers into four distinct groups: Shell, Services, Scenery and Set. The Shell is the structure; the Services covers all the mechanical and electrical as well as the water systems; the Scenery represents the internal layers including partitions, raised floors and ceiling and lastly the Set represents the fit out (Brand, 1997). Each one of these groups enters a process of decay with a varying rate that is not only problematic in terms of the close relationship of each layer to the rest and the difficulties it causes when the need to extract the redundant parts arises, but it also exhibits a dependence rather than an enriching relationship. The different layers provide the necessary rigidity, and thermal comfort and protection from the elements that make the building habitable, yet we are faced with the reality of a building process that extracts enormous amount of energy. This is followed by an equally energy-hungry operational life that leads to a gradual deterioration. Eventually the process culminates in obsolescence of materials that are difficult to reclaim or upcycle and that remain as waste products in the natural environment. This process of waste accumulation is particularly true to the human species. Unlike animal habitats found in nature, such as nests or burrowed caves, the building realm and mechanical solutions have a lasting impact on the environment. Our contemporary mode of construction rarely utilizes available materials and structures and cannot be absorbed back into nature, be broken down to feed organisms or provide aggregate. Human habitats are a product of irreversible energy-intensive processes, which is in stark contrast to vernacular construction that employs natural materials of the immediate surroundings (Stefanova *et al.*, 2019; Stefanova, 2021a).

However, Duffy's model of the building is particularly relevant in the context of biomaterials where the end product is oftentimes less permanent. As in the case of traditional construction, bio-materials, cannot be envisioned within a mono-material approach, rather they offer a combination of various layers that serve distinct functions. If we follow this logic, it is unreasonable to assume that a building would be entirely made of mycelium, kelp or biomineralized aggregate. Instead, it is more prudent to continue to view the building fabric as a carefully choreographed assemblage of parts, where different materials serve a purpose based on their material properties. Replacing Styrofoam, for example,which takes over 500 years to decompose with mycelium panels that can be degraded in months opens up the discussion of phasing out certain materials as alternatives become available. This also brings about questions of lifespans and conditions necessary for this type of construction. As designers, we are often faced with questions pertaining to the durability of biomaterials and the limiting factors when integrating them within traditional construction. However, if we examine traditional construction, accommodations are made for materials to be protected from moisture, such as cavity insulation, steel structural members, timber frames and electrical wiring, to name a few. If we apply this traditional lens to emerging materials, their integration in certain cases may simply be a question of replacing existing carbon positive materials with superior products.

It isn't until we start to conceive the building fabric as a living component that we enter the realm of speculation and this is where

Salvador Dali's famous statement comes to mind "The future of architecture will be soft and hairy". While we could continue to look at the building fabric as a static mass that enters a process of degradation as soon as it is formed, an alternative view could be taken, where the fabric assumes the role of an organism with its own metabolism that uses human waste products from its environment and releases a desirable output (Armstrong, 2015). In this scenario, we make allowances for fragile, wet, living, growing and dying components.

This type of construction would not imitate natural organisms, rather it would integrate them back into our domain not through their by-products or dead matter. In this scenario, the living organism executes a function closer to those we associate with mechanical systems, in the sense that bioremediation can occur on-site. This is not a mechanistic analogy, but rather it is the suggestion that the mechanical can be phased out by the biological and the self-generating. This later version does not depict an autonomous element functioning in isolation, but instead it is proposing continuous cycles and new networks between humanity and other species. Maintenance and care come to mind when thinking of living materials, which from a design perspective poses challenges. Our closest analogy is the integration of vegetation into buildings in the form of landscaping, green façades and roofs, which have promoted complex probiotic environments within urban and building settings. Widely employed examples include climbing plants or support systems that integrate living plants, along with a growing medium (soil or liquid/gaseous mixture of nutrients in the case of hydroponics) and a water distribution system (Manso and Castro-Gomes, 2015).

However, the microbial world comes with social barriers constructed by contemporary society. Inevitably fungus, bacteria and plankton are predominantly captured by mainstream media through the prism of pathogens and miasmas. This is in stark contrast to sustainability and regeneration that underly emerging practices. To socialize less widely accepted integration of microbial life into our built environment, experimental fabrication practices come to the forefront. Such work suggests the integration of living organisms and natural processes into the building realm, as illustrated in the work of Philip Beesley's Hylozoic Ground sculptural environments, which include dynamic processes and "living" technologies (Beasely, 2010). This experimental approach allows for emerging relationships to be tested in installations, prototypes and early adaptor technologies. We can begin to imagine a connection resembling the symbiotic relationship between the numerous healthy bacteria present in our body, for example. Suggestions of symbiotic relationships with living organisms expand human definitions of microbial life and open discussions as to the processes that could aid our faculties, offer us more acute sensing capabilities, improve management of our environment and create a bridge back to a closer coexistence with nature.

BIBLIOGRAPHY

Adamatzky, A., Ayres, P., *et al.* (2022) 'Logics in fungal mycelium networks', *Logica Universalis* [Preprint]. Available at: https://doi.org/10.1007/s11787-022-00318-4.

Adamatzky, A., Goles, E., *et al.* (2022) 'On fungal automata', pp. 455–483. Available at: https://doi.org/10.1007/978-3-030-92551-2-25.

Adamatzky, A. and Gandia, A. (2022) 'Living mycelium composites discern weights via patterns of electrical activity', *Journal of Bioresources and Bioproducts*, 7(1), pp. 26–32. Available at: https://doi.org/10.1016/J.JOBAB.2021.09.003.

Amyes, S.G.B. (2013) *Bacteria: A Very Short Introduction*. OUP Oxford (Very Short Introductions).

Antonelli, P. (2020) *The Neri Oxman Material Ecology Catalogue*. Edited by G.D. Lowry *et al.* New York: Museum of Modern Art.

Armstrong, R. (2015) *Vibrant Architecture: Matter as a Codesigner of Living Structures*. 1st edition. Warsaw: De Gruyter Open.

Armstrong, R. (2018) *Soft Living Architecture: An Alternative View of Bio-informed Practice*. 1st edition. London: Boomsbury Visual Arts.

Arnardottir, T. *et al.* (2020) *Turbulent Casting Bacterial Expression in Mineralized Structures*. ACADIA.

Bacon, F. (1626) *New Atlantis*. https://www.gutenberg.org/files/2434/2434-h/2434-h.htm

Battistuzzi, F.U., Feijao, A. and Hedges, S.B. (2004) 'A genomic timescale of prokaryote evolution: insights into the origin of methanogenesis, phototrophy, and the colonization of land', *BMC Evolutionary Biology*, 4(1), p. 44. Available at: https://doi.org/10.1186/1471-2148-4-44.

Beckwith, A.L., Borenstein, J.T. and Velásquez-García, L.F. (2022) 'Physical, mechanical, and microstructural characterization of novel, 3D-printed, tunable, lab-grown plant materials generated from Zinnia elegans cell cultures', *Materials Today*, 54, pp. 27–41. Available at: https://doi.org/https://doi.org/10.1016/j.mattod.2022.02.012.

Best, S. (2009) 'Genetic science, animal exploitation, and the challenge for democracy', in C. Gigliotti (ed.) *Leonardo's Choice*. Dordrecht: Springer Netherlands, pp. 3–19. Available at: https://doi.org/10.1007/978-90-481-2479-4-1.

Bierend, D. (2021) *In Search of Mycotopia: Citizen Science, Fungi Fanatics, and the Untapped Potential of Mushrooms*. London: Chelsea Green Publishing.

Birke, L. (2009) 'Meddling with medusa: on genetic manipulation, art and animals', in *Leonardo's Choice*. Dordrecht: Springer Netherlands, pp. 107–121. Available at: https://doi.org/10.1007/978-90-481-2479-4-7.

BIS (2010) *Estimating the Amount of CO2 Emissions that the Construction Industry Can Influence: Supporting Material for the Low Carbon Construction IGT Report*. London.

Bouillon, L. (2012) 'Ecological paint composition based on algae or algae extracts'. FELOR.

Braham, W.W., Hale, J.A. and Sadar, J.S. (2007) *Rethinking Technology: A Reader in Architectural Theory*. London: Routledge.

Brand, S. (1997) *How Buildings Learn: What Happens after they're Built*. Phoenix Illustrated.

Brown, J.R. and Doolittle, W.F. (1997) 'Archaea and the prokaryote-to-eukaryote transition', *Microbiology and Molecular Biology Reviews*, 61(4), pp. 456–502. Available at: https://doi.org/10.1128/mmbr.61.4.456-502.1997.

Cairns, S. and Jacobs, J.M. (2014) *Buildings Must Die: A Perverse View of Architecture*. Cambridge: The MIT Press.

Catts, O. and Zurr, I. (2004) *Victemless Leather, Tissue Culture and Art Project*. Available at: https://www.tca.uwa.edu.au/vl/vl.html (Accessed: 30 November 2020).

Chebath-Taub, D. *et al.* (2012) 'Influence of blue light on Streptococcus mutans re-organization in biofilm', *Journal of Photochemistry and Photobiology B: Biology*, 116, pp. 75–78. Available at: https://doi.org/10.1016/j.jphotobiol.2012.08.004.

Collet, C. (2020) 'Designing our future bio-materiality', *AI & SOCIETY* [Preprint]. Available at: https://doi.org/10.1007/s00146-020-01013-y.

Colmo, C. and Ayres, P. (2022) 'Remediating architecture: A bio-hybrid approach employing fungal mycelium', in *Structures and Architecture A Viable Urban Perspective?* London: CRC Press, pp. 27–34. Available at: https://doi.org/10.1201/9781003023555-4.

Crawford, A. (2022) 'Biodesign research in the anthropocene', in A. Mason and A. Sharr (eds) *Creative Practice Inquiry in Architecture*. 1st edition. London: Routledge. Available at: https://doi.org/10.4324/9781003174295.

Dessi-Olive, J. and Hsu, T. (2021) 'A design framework for absorption and diffusion panels with sustainable materials', *INTER-NOISE and NOISE-CON Congress and Conference Proceedings*, 263(4), pp. 2207–2218. Available at: https://doi.org/10.3397/IN-2021-2074.

Diamond v. Chakrabarty (1980) *Supreme Court of United States*. Available at: https://h2o.law.harvard.edu/cases/2017 (Accessed: 20 July 2021).

Dick, A., Bhandari, B. and Prakash, S. (2019) '3D printing of meat', *Meat Science*, 153, pp. 35–44. Available at: https://doi.org/10.1016/j.meatsci.2019.03.005.

van Dongen, T. (2015) *One Luminous Dot, teresavandongen.com*. Available at: https://www.teresavandongen.com/One-Luminous-Dot (Accessed: 14 December 2022).

EcoLogicStudio (2021) *EcoLogicStudio, ecoLogicStudio*. Available at: https://www.ecologicstudio.com/ (Accessed: 21 March 2022).

Elisar, O. (2018) 'The living language project'. Available at: https://doi.org/10.14236/ewic/EVAC18.42.

Elsacker, E. *et al.* (2019) 'Mechanical, physical and chemical characterisation of mycelium-based composites with different types of lignocellulosic substrates', *PLOS ONE*. Edited by D. Aydemir, 14(7), p. e0213954. Available at: https://doi.org/10.1371/journal.pone.0213954.

Elsacker, E., Søndergaard, A., *et al.* (2021) 'Growing living and multifunctional mycelium composites for large-scale formwork applications using robotic abrasive wire-cutting', *Construction and Building Materials*, 283, pp. 122732. Available at: https://doi.org/10.1016/j.conbuildmat.2021.122732.

Elsacker, E., Vandelook, S., *et al.* (2021) 'Mechanical characteristics of bacterial cellulose-reinforced mycelium composite materials', *Fungal Biology and Biotechnology*, 8(1). Available at: https://doi.org/10.1186/S40694-021-00125-4.

Fairs, M. (2013) *Mycelium Chair by Eric Klarenbeek is 3D-printed with Living Fungus, Dezeen*. Available at: https://www.dezeen.com/2013/10/20/mycelium-chair-by-eric-klarenbeek-is-3d-printed-with-living-fungus/ (Accessed: 22 October 2020).

Fisher, D. (2021) *Self-Healing Concrete*. Millersville: Materials Research Forum LLC (Materials Research Foundations). Available at: https://doi.org/10.21741/9781644901373.

Furusawa, C. and Kaneko, K. (2002) 'Origin of multicellular organisms as an inevitable consequence of dynamical systems', *The Anatomical Record*, 268(3), pp. 327–342. Available at: https://doi.org/10.1002/ar.10164.

Gnanam, C. (2019) *Introduction to Mycology*. 1st edition. Triplicane: MJP Publishers.

Haraway, D.J. (2016) *Staying with the Trouble: Making Kin in the Chthulucene*. Durham: Duke University Press.

Hawksworth, D.L. (2001) 'The magnitude of fungal diversity: the 1.5 million species estimate revisited', *Mycological Research*, 105(12), pp. 1422–1432. Available at: https://doi.org/10.1017/S0953756201004725.

Heisel, F. *et al.* (2017) 'Design of a load-bearing mycelium structure through informed structural engineering', In *Proceedings of the World Congress on Sustainable Technologies (WCST)* (pp. 1-5).

Heveran, C.M. *et al.* (2020) 'Biomineralization and successive regeneration of engineered living building materials', *Matter*, 2(2), pp. 481–494. Available at: https://doi.org/10.1016/j.matt.2019.11.016.

Hird, M.J. (2009) *The Origins of Sociable Life: Evolution after Science Studies.* New York: Palgrave Macmillan.

Ilgun, A. and Schmickl, T. (2022) 'Mycelial beehives of HIVEOPOLIS: designing and building therapeutic inner nest environments for honeybees', *Biomimetics*, 7(2), p. 75. Available at: https://doi.org/10.3390/biomimetics7020075.

Karana, E., Pedgley, O. and Rognoli, V. (eds) (2014) *Materials Experience: Fundamentals of Materials and Design.* Oxford: Elsevier.

Kavanagh, K. (ed.) (2017) *Fungi: Biology and Applications.* Hoboken, NJ: John Wiley & Sons, Inc. Available at: https://doi.org/10.1002/9781119374312.

Keel Labs (2022) *Kelsun, Keel Labs.* Available at: https://www.keellabs.com/kelsun (Accessed: 19 December 2022).

Keune, S. (2021) 'Designing and living with organisms weaving entangled worlds as doing multispecies philosophy', *Journal of Textile Design Research and Practice*, 9(1), pp. 9–30. Available at: https://doi.org/10.1080/20511787.2021.1912897.

Kim, K.H. (2022) *Microalgae Building Enclosures: Design and Engineering Principles.* New York: Taylor \& Francis. Available at: https://books.google.com/books?id=-JFfEAAAQBAJ.

Kirk, P.M. *et al.* (2001) *Ainsworth & Bisby's Dictionary of the Fungi.* 9th edition. Oxford: Oxford University Press.

Lorimer, J. (2020) *The Probiotic Planet.* London: University of Minnesota Press. Available at: https://doi.org/10.5749/j.ctv19cw9vg.

Lowenfels, J. and Lewis, W. (2006) *Teaming with Microbes: A Gardener's Guide to the Soil Food Web.* London: Timber Press.

Maheshwari, R. (2016) *Fungi: Experimental Methods in Biology.* 2nd edition. Boca Raton: CRC Press.

Mandal, S. *et al.* (2023) 'Algal polysaccharides for 3D printing: a review', *Carbohydrate Polymers*, 300. Available at: https://doi.org/10.1016/J.CARBPOL.2022.120267.

Manso, M. and Castro-Gomes, J. (2015) 'Green wall systems: a review of their characteristics', *Renewable and Sustainable Energy Reviews*, 41, pp. 863–871. Available at: https://doi.org/10.1016/j.rser.2014.07.203.

McCoy, P. (2016) *Radical Mycology: A Treatise on Seeing and Working with Fungi.* Portland: Chthaeus Press.

Merchant, C. (1980) *The Death of Nature: Women, Ecology, and the Scientific Revolution.* 1st edition. San Francisco: Harper and Row.

Mohammadifar, M. *et al.* (2018) 'Power-on-paper: origami-inspired fabrication of 3-D microbial fuel cells', *Renewable Energy*, 118, pp. 695–700. Available at: https://doi.org/10.1016/j.renene.2017.11.059.

Muthu, S.S. and Rathinamoorthy, R. (2021) *Bacterial Cellulose: Sustainable Material for Textiles*. Singapore: Springer Singapore (Sustainable Textiles: Production, Processing, Manufacturing /& Chemistry).

Myers, W. and Antonelli, P. (2014) *Bio Design: Nature, Science, Creativity*. London: Thames and Hudson Ltd.

Myers, W. and Antonelli, P. (2018) *Bio Design: Nature, Science, Creativity*. Revised edition. London: Thames and Hudson Ltd.

Ngan, N.D. *et al.* (2019) 'Tissue-cultured human cord lining epithelial cells in treatment of persistent corneal epithelial defect', *Open Access Macedonian Journal of Medical Sciences*, 7(24), pp. 4266–4271. Available at: https://doi.org/10.3889/oamjms.2019.372.

Ozkan, D. *et al.* (2022) 'Are mushrooms parametric?', *Biomimetics*, 7(2), p. 60. Available at: https://doi.org/10.3390/BIOMIMETICS7020060.

Parhizkar, H. and Afghani Khoraskani, R. (2017) *Green facade system for indoor air purification*. In *12th Conference on Advanced Building Skins*. Bern: Advanced Building Skins GmbH, pp. 1151–1160.

Pawlyn, M. (2016) *Biomimicry in Architecture*. 2nd edition. London: RIBA Publishing.

Pedgley, O., Rognoli, V. and Karana, E. (eds) (2021) *Materials Experience 2: Expanding Territories of Materials and Design*. Cambridge: Elsevier. Available at: https://doi.org/10.1016/C2018-0-03833-5.

Reynolds, C. (2006) *Ecology of Phytoplankton*. 1st edition. Cambridge: Cambridge University Press.

Rigobello, A., Gaudillière-Jami, N. and Ayres, P. (2022) 'Prototaxites stellaviatori: a fungal growth simulation model for Mycelium-Based Composites education in applied arts', in *Structures and Architecture A Viable Urban Perspective?* London: CRC Press, pp. 35–42. Available at: https://doi.org/10.1201/9781003023555-5.

Roosth, S. (2017) *Synthetic: How Life Got Made*. Chicago: University of Chicago Press.

Ross, P. (2019) 'Monokaryon mycelial material and related method of production'. MycoWorks, Inc.

Santhanam, R. (2022) *Bioluminescent Marine Plankton*. BENTHAM SCIENCE PUBLISHERS. Available at: https://doi.org/10.2174/978981 50502021220101.

Sawa, M. (2012) 'Algaerium', in W. Myers (ed.) *Bio Design: Nature: Science Creativity*. London: Thames Hudson, pp. 88–91.

Schaffner, M. *et al.* (2017) '3D printing of bacteria into functional complex materials', *Science Advances*, 3(12), p. eaao6804. Available at: https://doi.org/10.1126/sciadv.aao6804.

Senarathna, S., Wickramasinghe, I. and Navaratne, S. (2022) 'Current applications of seaweed-based polysaccharides in edible packaging', *Algal Functional Foods and Nutraceuticals: Benefits, Opportunities, and Challenges*, pp. 447–464. Available at: https://doi.org/10.2174/9789815051872122010022.

Sheldrake, M. (2020) *Entangled Life: How Fungi Make Our Worlds, Change Our Minds and Shape Our Futures.* London: Bodley Head.

Stamets, P. (2005) *Mycelium Running: How Mushrooms Can Help Save the World.* New York: Ten Speed Press.

Stamets, P. and Chilton, J.S. (1983) *The Mushroom Cultivator: A Practical Guide to Growing Mushrooms at Home.* Seattle: Agarikon Press.

Stefanova, A. *et al.* (2019) 'Approach to biologically made materials and advanced fabrication practices', in M. Asefi and M. Gorgolewski (eds) *International Conference on Emerging Technologies In Architectural Design (ICETAD2019).* Toronto: Ryerson University, pp. 193–200.

Stefanova, A. (2021a) 'Beyond biomimicry: developing a living building realm for a post-anthropocene era', in O. Pedgley, V. Rognoli, and E. Karana (eds) *Materials Experience 2: Expanding Territories of Materials and Design.* Oxford: Butterworth-Heinemann.

Stefanova, A. *et al.* (2021) 'Photosynthetic textile biocomposites: using laboratory testing and digital fabrication to develop flexible living building materials', *Science and Engineering of Composite Materials*, 28(1), pp. 223–236. Available at: https://doi.org/10.1515/secm-2021-0023.

Stefanova, A. (2021b) 'Towards a post-anthropocene bio-design practice', in A. Sharag-Eldin and C. Jarrett (eds) *ARCC 2021: Performative Environments.* Tucson: Architectural Research Centers Consortium, Inc., pp. 291-.

Tandon, N. and Joachim, M. (2014) *Super Cells: Building with Biology.* TED Conferences. Available at: https://www.amazon.com/kindle/dp/B00ICXN3VA/ref=rdr_kindle_ext_eos_detail (Accessed: 29 November 2020).

The Human Microbiome Project (2012) 'Structure, function and diversity of the healthy human microbiome', *Nature*, 486(7402), pp. 207–214. Available at: https://doi.org/10.1038/nature11234.

Tostões, A. (2020) 'Health at the core of modern movement architecture', *Cure and Care*, 62, pp. 2–3. Available at: https://doi.org/10.52200/62.A.6QVKSDMB.

Tsing, A. (2017) *The Mushroom at the End of the World: On the Possibility of Life in Capitalist Ruins.* Princeton: Princeton University Press.

Xing, Y. *et al.* (2018) 'Growing and testing mycelium bricks as building insulation materials', in *IOP Conference Series: Earth and Environmental Science.* Institute of Physics Publishing. Available at: https://doi.org/10.1088/1755-1315/121/2/022032.

Yong, E. (2016) *I Contain Multitudes: The Microbes Within Us and a Grander View of Life.* New York: Random House.

Design in the Lab

Chapter 2

DOI: 10.4324/9781003363774-2

APPROACHES IN BIO-DESIGN

In architecture, we have become accustomed to the growing rift between design and making. The design world has transitioned from the medieval ideal of the master builder, the craftsperson and maker into a distanced mode of abstraction and instruction through representation (Ingold, 2000). The making process therefore has become a foreign domain and assigned a mechanical role on the way to an outcome. This has led to a push for standardization and predictability of performance. This mode of creating is heavily reliant on inanimate, standard components that exhibit predictable and stable behavior. Therefore, the introduction of living materials into everyday environments calls for a new mode of design that is guided by the lifecycle of materials and the material development and design of support systems for microbial life. The former can be labeled as top-down design, a mode of design that looks at the broader integration and interaction of living materials, a process more akin to traditional architectural design practice (Dade-Robertson, 2020). The latter option, therefore, can be viewed as bottom-up design, which originates from the needs of the living organisms, a significantly less deterministic method of working.

Traditional top-down approaches to bio-design are grounded in technocratic thinking typically present in engineering and computer science (Ramirez Figueroa, 2018). Therefore, we often encounter reductive metaphors that seek to standardize self-sustaining elements that designers may wish to incorporate in a similar way to inanimate design components (Stefanova, 2021b). However, living matter presents a greater level of complexity in its needs and interaction with the environment and the relationship between the human inhabitants and the living species designers seek to incorporate into human settings. Hence, top-down bio-design is riddled with pitfalls pertaining to assumptions, oversimplification, unreasonable choreography of human and nonhuman entities, as well as the urge to resort to biotechnical interpretations of the living building realm.

HOME LABS AND EXPERIMENTAL DESIGN SPACES

There is a growing bio-design community that practices from domestic to design studio spaces. These practices have brought about a revolution of new bio-products, sustainable design and

Design in the Lab

Figure 2.1 Bio-designer Dilan Ozkan's experimental home setup during the COVID-19 lockdown, 2020 [Image by Dilan Ozkan].

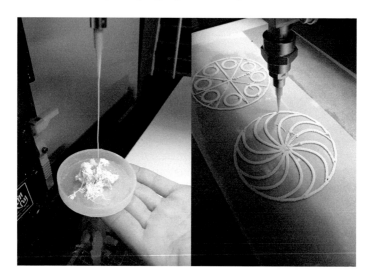

Figure 2.2 Using a clay 3D printer to extrude algae-laden matrix, work conducted in the workshop of the Architecture Department of Newcastle University, 2021.

material research with direct design applications. Such endeavors were amplified by the COVID-19 pandemic, during which time we saw a migration of lab experiments from formal academic settings to unconventional home setups. These nonspecialist spaces rely on the ingenuity and resourcefulness of the practitioner, sometimes limiting the types of explorations that can take place as well as the methods employed. Often, this also limits the types of organisms that can be studied due to availability as well as the regulatory restrictions surrounding certain species (Figure 2.1).

There are various examples of scientific work taking place outside the lab and in domestic or maker settings that benefit from the ability to scale and fabricate with living organisms using tools not typically associated with traditional laboratories. These spaces present challenges in accessing the necessary tools to ensure the same level of precision associated with scientific lab practice or maintain the high level of cleanliness that reduces the chances of contamination (Figure 2.2).

WORKING IN A SCIENCE LAB: SHADOWING, COLLABORATION AND CONVERSATION

Bio-design is still a challenging field to enter due to its interdisciplinary nature, where specialists are individually responsible for establishing relationships across various disciplines, which is a challenging first step due to the lack of mechanisms within academic or private research settings for those collaborations to occur. Research centers have recently been established for the particular purpose of offering researchers an avenue into bio-design practice. Examples of such collaborative initiatives include the Hub for Biotechnology in the Built Environment in Newcastle upon Tyne, Bio-Integrated Design (Bio-ID) Lab in London, Centre for Information Technology and Architecture (CITA) in Copenhagen and the Department of Design Tech at Cornell University. However, such examples are often reliant on short-term funding, making the functions of such groups precarious and require a large initial outlay to set up professional networks, source equipment and spaces for such work. In certain cases, this initial groundwork alone takes a large proportion of the project lifetime just to offer designers the much needed bridge between disciplines, subsequently placing considerable financial burdens upon the project.

Still, the more common model for design practitioners of establishing collaborations with scientific partners through personal efforts remains. This may entail reaching out to various scientific

Design in the Lab

groups to establish potential synergies in objectives where a designer may be given access to research facilities. In the event of a new practitioner entering the lab, the residency is most likely to begin with various discussions as well as a shadowing period. This is an invaluable acquisition of skills and knowledge and is often a learning model for early-stage science students as well as specialists from other fields. Although many programs are structured around short-term visits and limited time within laboratory settings, effective uptake of methods and development of deeply embedded practice in my experience requires longer periods of emersion. The design of scientific research spaces, as pointed out by Marin Sawa (Sawa, 2016), is particularly conducive to organic exchanges of knowledge and ideas through conversations that occur due to densely populated spaces and long hours spent in the lab. These exchanges are invaluable in understanding processes, learning scientific jargon and becoming familiar with lab practice. These collaborative modes of coworking grow over time and new common points of interest are naturally established along the way as trust and respect grow among the working group (Stefanova, 2021a). The initial barriers of scientific and design philosophies begin to spark new opportunities, opening avenues into previously inaccessible ground.

For an equal exchange to occur, each practitioner must bring a set of expertise cultivated within a particular discipline, for instance, a foundation in a design field such as architecture, which later may serve as a resource within a multidisciplinary project, where every member of the team's expertise provides a unique lens for exploration. Identification of mutually beneficial goals can also facilitate a closer collaborative environment. As science is largely disseminated through academic publication and outputs with an associated metric, designers can contribute to scientific teams by adopting this mode of publication. It is also worth noting that respecting intellectual property and crediting the scientific team is imperative for long-term working relations. Design fields often celebrate the sole practitioner, with various pieces of writing credited to the individuals directly executing the work or participating in the writing process. Within the sciences, publications often list multiple authors that include the people executing the work, people who have supervised the work such as lab leaders, people who have intellectually contributed to the work through development of adopted protocols and/or people who have formally or informally offered advice. This long string of authors does not take away from the work, but rather it underlines a healthy work culture and a rich collaborative environment.

Another aspect of lab practice is the associated cost of conducting this type of research, in particular consumables, equipment and the cost of running tests using outside instruments. Therefore, an important part of working and collaborating with scientific partners is sourcing funding that enables scientists to dedicate the needed time to the project. Some of the funding bodies that regularly advertise funding applicable to the field of bio-design include UK Research and Innovation (UKRI), European Research Council (ERC), Australian Research Council (ARC) and US funding bodies such as the National Endowment for the Humanities (NEH), National Science Foundation (NSF), National Aeronautics and Space Administration (NASA), Environmental Protection Agency (EPA) and National Endowment for the Arts (NEA).

HEALTH, SAFETY AND WASTE DISPOSAL
In addition to procedures directly related to the particular task you are executing, many of the shared laboratory spaces, especially those belonging to established institutions, will have their own health and safety standards needing to be followed, which may involve paperwork, chemical storage, dress code, labeling and storage of samples, disposal of chemicals and specimens and cleaning procedures before and after the use of spaces and instruments (Gamborg and Phillips, 1995). It is important to adhere to such norms and respect the processes in place, as they have been established to protect both you and your work as well as the work of others. For example, in several labs that I have worked in, multiple organisms and various types of work have taken place at the same time. Outbreaks of contamination occasionally occurred within those spaces either when procedures did not make provisions for potential contamination risks or when procedures were not followed. This often resulted in considerable disruptions to work, due to the lab needing to be thoroughly disinfected, as well as, on occasions, loss of work conducted over multiple months, creating considerable setbacks for those involved. There are also more serious events that could occur, where the health and safety of the people working within those spaces could be put at risk. If you operate from a home or a studio environment, it is also essential to dispose of waste

Design in the Lab 27

safely for the aforementioned reasons. A sharps bin and personal protective equipment are the most basic and essential items to consider, along with the safe storage of samples, which should be stored within a separate refrigerator designated for biological substances. Any biological work should also be accompanied by research into potential risks of using the chosen living organisms. Even the most common natural organisms may present dangers to human health and thus may require a guide to working practices.

Designers entering the world of lab practice sometimes view such health and safety regulations as a nuisance, unnecessary obstacles to making or creativity, which resulted in some of the greatest sources of friction I have witnessed in scientific spaces between designers and scientists. During my early days in the lab, I decided to view such procedures as choreography that I agreed to follow out of respect for other people, their spaces and their equipment. I also appreciated that the prescribed procedures were there for reasons that may not have been apparent to me at that time. The rationale behind these foreign health and safety measures became clearer, the longer I spent in these shared spaces and even more so now that I run my own lab.

SETTING UP A DESIGN WET LAB
Setting up a lab space requires forward planning in terms of establishing the type of work that will be conducted within the space, the species used, the number of people using the space, maintenance and safety procedures. It is unlikely that you will be able to source or plan for every possible type of work you may wish to conduct within the space in the future, but certain aspects would have to be considered in most instances, including storage and disposal of chemicals and biological agents, cleaning, maintenance, health and safety protocols and schedules. Another thing to consider is the space occupancy and the available area, including the optimal layout that is related to workflows.

Some of the essential equipment and measures you need to consider include access control, personal protective equipment and running water, with the addition of deionized water filters, to name a few. This is likely to be supplemented by an area for aseptic work in the form of a laminar flow hood, monotub or a Bunsen burner, discussed further later in the chapter. For sterilizing equipment, media and other substrates, you will need either an autoclave or a pressure cooker. Some of the first items worth sourcing may be a fridge not used for food purposes, a flammables cabinet and a sharps bin. These will also have to be accompanied by a lab management plan that establishes disposal protocols of lab waste. If the lab is going to be a part of an institution for higher education, it is likely that there are regulations and processes in place needing to be followed. However, if the lab functions independently or within a department not typically associated with the sciences, you may have to put those measures in place independently. Based on the type of work conducted within the space, the lab is going to have a lab safety level or containment level that will dictate the safety measures needing to be in place. You can find those requirements from governing bodies in your country of operation, for example, the Centre for Disease Control and Prevention in the US (CDC, 2021) or the Health and Safety Executive in the UK (ACDP, 2019). These governing bodies often provide the legal requirements as well as advice for best practices and lab design.

As the facilities and equipment are largely going to depend on the type of work conducted within the space, it is essential to survey similar facilities and literature that outlines various requirements to establish the initial list. Refer to Table 2.1 for a basic list of equipment and consumables.

Table 2.1 Consumables and Equipment

This table is not an exhaustive list of everything you may need or may be required to have but a suggested list of some of the equipment and consumables for starting a lab that handles microorganisms that are not a danger to human health (e.g., Containment Levels 1 and 2 or Biosafety Lab Levels 1 and 2). For an up-to-date and comprehensive list of all requirements, refer to appropriate government agency

General list of consumables and equipment
• Fridge freezer (for biological work) • Flammables cabinet • Sink • Deionized water system • Autoclave/pressure cooker/Bunsen burner • Protective equipment (disposable gloves, lab coat, goggles, etc.) • Sharps bin • Scales (recommended to have three to four decimal places depending on the level of precision)

(*Continued*)

Design in the Lab   29

Table 2.1 (Continued)

- Glassware (that can withstand pressure and heat)
- Spatulas, tweezers and small metal tools
- Sterile incubation vessels (petri dishes/well plates/tubes, etc.)
- Parafilm
- Autoclave tape
- Microscope with various objective lenses, glass slides and glass slips
- Magnetic stirrer
- Hot plate
- Pipettes (200 µl and 1,000 µl) and pipette tips
- pH meter
- 70% Ethanol or isopropyl alcohol
- Cleaning products and tools

Space management measures

- Access control
- Waste management plan
- Maintenance schedule
- Health and safety file and other relevant documentation and warning signage depending on the country and the facility type and location

Species-specific equipment and consumables

Algae	Fungus	Bacterial cellulose
<ul><li>Centrifuge</li><li>Falcon tubes/ epidorphs</li><li>Culture incubation containers (e.g., clear carboys, large flasks, glass media bottles)</li><li>Multi-port lids</li><li>Air filters</li><li>Plastic tubing</li><li>Air pump</li><li>Lights on a timer</li><li>Hygrometer</li><li>Incubation area</li></ul>	<ul><li>Mushroom bags</li><li>Humidifier</li><li>Dark incubation tent</li><li>Heater that can handstand high moisture environments</li><li>Hygrometer</li><li>Large gage needles</li><li>Syringes of various sizes</li></ul>	<ul><li>Dark incubation area with controlled temperature or incubator</li><li>Growth vessels depending on the type of experiment (large plastic tubs, six-well plates, etc.)</li><li>pH meter</li></ul>

Depending on the type of lab, you may be able to source a large proportion of the consumables or smaller pieces of equipment from general, nonscientific suppliers or purchase secondhand equipment. Many higher education institutions also dispose of functional equipment that they are willing to donate or sell at a nominal price.

CONTAMINATION AND ASEPTIC CONDITIONS

What is commonly considered clean within an everyday setting is in fact already a home to spores, bacteria, fungi, algae, etc. There are a few ways to create a sterile space, where you can safely work and prevent most contamination. Such methods include using a Bunsen burner, a laminar flow hood or a monotub. The use of an open flame, typically from a Bunsen burner, can offer a sterile zone below the flame of approximately 15 cm, depending on flame size and air movement. This zone offers an aseptic working area where various types of aseptic transfers of living material and nutrients can occur. The heat from the flame creates a convection current that pushes particles upward and away from that zone (Sanders, 2012a).The size of this space prohibits work with larger containers; however, it is an excellent starting point that offers practitioners flexibility when working in spaces that don't offer laminar flow hood facilities. Some good guides to aseptic work with a pipette is "Aseptic Laboratory Techniques: Volume Transfers with Serological Pipettes and Micropipettors" and "Aseptic Laboratory Techniques: Plating Methods" by Erin Sanders (Sanders, 2012a, 2012b) that outline aseptic transfer and inoculation with a Bunsen burner (Figure 2.3).

The monotub method based on a glove box or still air box design is another relatively simple, but effective way for facilitating aseptic work. Again it is limited by the size of the container used and requires a clear plastic tub to offer visibility and create a still air working zone that minimizes the chances of contamination. To make such a container, you would need to make two circular openings into the side of the tub. The easiest way to do this is by heating an empty tin can with a heat gun and pushing it through the plastic. When using the tub, all the working materials must be placed underneath prior to starting the procedure, including a bottle of 70% alcohol. Use nitrile gloves and disposable arm protectors. Place your arms inside and generously spray the interior including your gloves with the

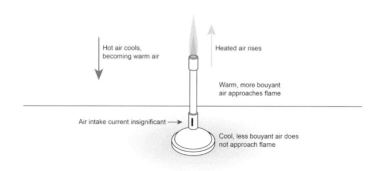

Figure 2.3 Sterile zone created by a Bunsen burner.

Figure 2.4 Still air box.

alcohol to kill any bacteria before proceeding to conduct the work (Kosel and Ropret, 2021). This is an appropriate option when work is to be conducted in settings other than a designated lab, such as workshops, studios or home environments (Figure 2.4).

Another option is using a laminar flow hood, either purchased off-the-shelf or self-constructed, as an ideal setup at low cost. Many labs will already have a laminar flow hood, and secondhand laminar hoods can be purchased at a reduced rate from various suppliers of refurbished lab equipment. It is also worth asking higher education institutions as they often discard or sell unwanted equipment at minimal rates. Another option is to construct a laminar flow box, which is standard practice within science and design settings (Meyer, 1986). To do this, you will need a high-efficiency particulate absorbing filter (HEPA) filter with 99.99% efficiency at 0.3 microns; a preferred size is 24" × 24" or larger to allow for a larger work area. These filters are typically thicker than a standard filter, and they are quite costly. The second piece of equipment is a fan that can achieve a velocity of 100 ft/min. This requires a calculation to find the right filter so you will need to know the resistance of the filter, which is usually printed on the product, which in many cases is 1.1–1.3 in. w.g. (inches of water gauge) (Shields, 2022).

$$\text{Volumetric flow rate} = \text{Velocity} \times \text{Area}$$

Example:

$$\text{Volumetric flow rate} = 100 \text{ ft} / \min \times (24 \text{ inch} \times 24 \text{ inch})$$

$$(24 \text{ inch} \times 24 \text{ inch} = 586 \text{ inch} = \text{apx. } 4 \text{ ft}^2)$$

$$= 100 \text{ ft} / \min \times (4 \text{ ft}^2)$$

$$\text{Volumetric flow rate} = 400 \text{ ft}^3$$

The above equation provides the required flow rate (EngineersEdge, 2022). However, the filter is going to have resistance that is going to have an impact on the flow rate. Therefore, when selecting a fan, you will need to look at the fan curve, which is normally supplied by the manufactures. This is a table that plots pressure against flow rate shown in Figure 2.5.

The box itself, normally constructed out of timber, should be approximately 35 cm deep and must encase the perimeter of the filter. A rectangular opening must be placed above to allow the fan to be mounted on top. A secondary box with a smaller filter should be placed over the fan to prevent dust particles from settling on top of

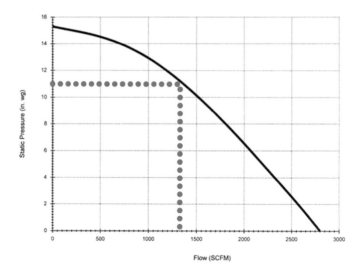

Figure 2.5 You need to locate the filter's resistance, look for the desired in. w.g. and draw a straight line until it intersects with the curve. The intersection point will give you the volumetric flow rate in relation to your filter.

the filter shown in Figure 2.6. This secondary filter can be a general-domestic filter.

Regardless of whether the laminar flow hood is purchased as an off-the-shelf product or an assembled box, the filters must be tested every two months to ensure that they are still functioning as expected. This can be done by leaving a petri dish with nutrients open within the negative pressure zone, approximately 10 cm away from the filter for 10 min and incubating for a few days to see if any bacterial contamination appears on the surface (Staat and Beakley, 1968). In the event this occurs, the HEPA filter must be replaced to prevent contamination.

In almost all instances, ethanol or isopropyl alcohol of no less than 70% should be used to disinfect all tools and surfaces that are being used. Disposable nitrile gloves and a protective lab coat and, in certain cases, disposable arm protectors should be used to protect both you and your work. It is worth noting that the level of sterility is often determined by the type of organisms used. For example, if working with algae, discussed later, you may not require the use

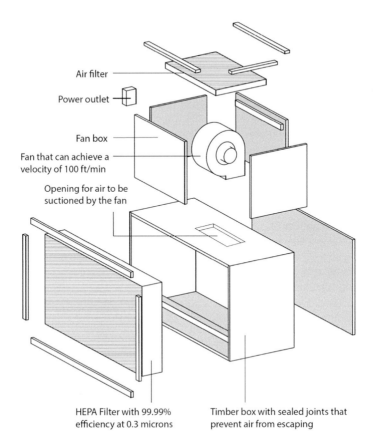

Figure 2.6 Laminar flow box construction.

of a laminar flow hood when working with liquid culture, provided you are using a single species within the space, and thus, potential contamination sources are reduced. However, when working with fungi, most cultivation steps must be performed within aseptic conditions, especially in the early stages, when the organism has not colonized large zones of the substrate. This factor may influence your decision to work with certain types of organisms due to the facilities and equipment at your disposal. There may also be occasions when the work involves genetically modified organisms that may have to be handled in accordance with bespoke protocols or organisms that

Design in the Lab

are harmful to humans, in which case a laboratory partner should be consulted to establish correct health and safety processes.

SPECIES MATTER

Species selection is perhaps one of the most important parts of designing your experimental study. It is a fundamental driver for the processes you employ and the outcomes you hope to achieve. The biodiversity on the planet offers the possibility for an unimaginable number of biological solutions to problems. However, the identification of the most appropriate species for the work you are trying to conduct becomes the challenging task. For example, there are microalgae that have superior carbon dioxide fixing capabilities; others are suited to extreme environmental conditions (extremophiles) such as pH and low or high temperatures. There are also those organisms that metabolize waste products, including heavy metals as well as chemical or biological waste; hence, they are bioremediators. Therefore, prior to initiating your research, again, it is vital to establish what you are trying to achieve and then conduct a literature review to establish the species commonly associated with the performative functions you wish to harness.

This initial investigation is likely to be accompanied by identification of other factors, which may include your ability to source and sustain the organism in question. For example, there are various fungal species that are mycorrhizal and therefore would not thrive in isolation within an *in vitro* environment. There are also species that may be recorded within the literature that you do not have access to within your research setting.

In addition, there is the design compatibility of the organism and the intended application. For example, if designing for a building setting, it may not be viable to include large liquid culture containers, or you may be limited by the amount of cultivation space needed to effectively remediate waste from a particular source. In essence, the choice of organism is the most important factor that will determine the trajectory of your research, and a large proportion of the initial work may constitute establishing that compatibility. This is done through studying several species and running comparative studies to establish their potential in relation to the problem you are trying to solve.

DESIGNING A PROCESS

Following a literature review, an experimental method needs to be developed and followed in order to obtain repeatable results and data that supports or rejects the hypothesis. The variables of the study need to be identified, and in the context of *in vitro* fundamental wet lab studies, all other variables must be controlled to establish if the identified variable has a bearing on the process and to what degree. The variables that are going to be studied are known as the "dependent variables" and those that are to be maintained as constant by the researcher are known as the "independent variables" (Bell, 2009). For instance, some variables you may wish to study are nutrient dilutions, incubation temperature, matrix composition, substrate compatibility with the identified organisms, pH levels, incubation conditions, etc. Once the variables are identified, there needs to be a precise plan that stipulates tools, reagents, exact amounts and incubation conditions such as time, light levels, light cycles, humidity and temperature. The sequence and a specific way of setting up the experiment must be established and may include conditions for growing the culture before the experiment, tools and methods for preparing the samples, such as mixing protocols, incubation containers and transfer technique, to name a few. It is important that variables are tested independently and then potentially combined to ascertain if there is an interaction between the two that is statistically significant.

Once the experiment is set up, data collection becomes of great importance. There are qualitative and quantitative types of data collection. Qualitative data is a type of data categorized through descriptive means, in contrast to quantitative data that focuses on numerical data. Qualitative data in the context of fundamental wet lab experiments may involve various observational methods such as photography, observation of cell behavior under a microscope, various spectroscopy studies and capturing of cell fluorescence. Quantitative analysis, on the other hand, may involve pH measurements, weighing of samples, material property testing to assess compressive and tensile strength, carbon dioxide levels, etc.

Once data is collected, it can be analyzed either through comparative observational means or through statistical analysis that considers the number of samples (n value) along with the standard deviation between replicates (s.d.), the data distribution that will indicate the type of statistical method that needs to be used based on whether or not the data is normally distributed. The data can be used to establish if the variable or the interaction between variables is statistically significant by obtaining a p-value. A $p < 0.05$ indicates a threshold of significance; therefore, the variable is statistically significant, whereas if $p > 0.05$, then the variable is not significant. Calculations to assess significance are usually completed using statistical software such as Minitab, Excel plug-ins, IBM SPSS Statistics and RStudio, to name a few.

Each aspect of conducting the work will be recorded in the "Method" section of a scientific paper, if that is an avenue you chose to utilize to disseminate your work. The method will also be beneficial in progressing the research by following and developing protocols that can serve as a basis for future work.

WORKING IN MULTIPLES, EXPERIMENTS AND SETTING UP CONTROLS

In the design field, there is a practicality that often relies on prototypes and simulation to establish the potential viability of outcomes. When it comes to wet lab experiments, a single successful outcome is not considered definitive proof. This is particularly true when working with living organisms as variations inevitably occur, even if the same protocol is followed. The variations can often be attributed to the behavior of the individual colonies, subtle variations within the incubation environment and human error or technique. To that end, scientific experiments seek to minimize these variations in several ways, including working in multiples. If you are to inoculate several agar plates at the same time with the same culture, you will observe very visible differences in the days that follow. Some plates may exhibit high colonization rates and other may lag. Therefore, when taking readings, conducting observational studies or trying to assess levels of success, it is essential to conduct experiments with multiple samples, inoculated at the same time and incubated under identical conditions. This is often done in triplicates, although the larger the number, the more reliable the experimental results become. To ensure repeatability and verify that human factors are unlikely to have influenced the results, it is also advisable to run experiments multiple times.

Another big factor is time and the age of the culture used. When conducting experiments, testing multiple strains, for example, including controls, all samples must be inoculated on the same day. This is largely to do with the age of the culture, which must be consistent, as it plays a vital role in the development of the sample. For example, if, in one instance, the algae culture you are using is two weeks old and the other culture is six weeks old, there is likely to be a difference in the cell density, the amount of dead cell matter and the nutrients within the culture, which in turn may influence the development of the cells. The optimal age of the culture is something to be established from the available literature, as that is likely to vary between species and strains.

This brings us to the role of controls within laboratory experiments, which, admittedly, were a big sticking point in my design geared practice. As an example, when I first entered the field, I struggled to understand why empty controls were necessary. If there is an absence of an organism, then surely nothing would happen. However, this is where the difference between assumption and empirical knowledge comes into play. To make that statement and to be completely sure that the observed results are not caused by the environment, a chemical reaction in the media, deterioration of the substrate, etc., it is essential to have physical proof. Therefore, when setting up experiments, there is usually a set of biological control samples incubated without the presence of the tested variable as well as a second set without the presence of the living organism, but including the variable to be assessed, shown in Figure 2.7.

VIEWING THE WORLD

The microbial world has been made increasingly more accessible through the development of modern microscopy and imaging tools that can capture properties that are not apparent in visible light (Fishel, 2017). In working with photosynthetic microorganisms, there are several options that become available, depending on the size of the cells. The same goes for fungal or bacterial samples, which may be suited to various magnification levels, types of analysis and methods. Here are some examples of tools that may be applicable in assessing various samples.

Imaging Larger Samples Areas

The stereo microscope allows samples to be imaged without prior preparation, by placing the whole petri dish on the podium of the microscope. These microscopes are typically of lower magnification,

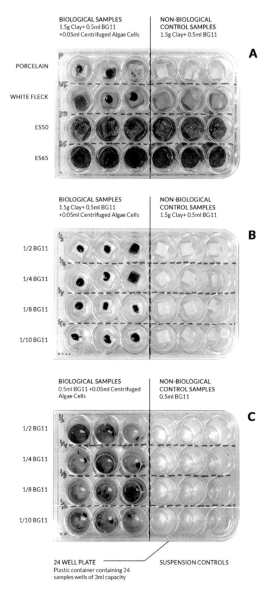

Figure 2.7 Microalgae cultivated in media of various dilutions on clay substrates along with biological and nonbiological controls (Stefanova *et al.*, 2020).

ranging between ×0.5 and ×60 (Li, 2013), and are better suited to work with larger samples, rather than single-cell analysis.

There are also advanced 3D digital microscopes based on a similar imaging setup as the traditional stereo microscope. They capture various planes at the same time and combine them so that all parts of the imaged area are in focus. This is particularly appropriate for fungal samples, where the hyphae form a mesh that sits on multiple planes.

Imaging Individual Cells

Another option is using a light microscope. The quality and capabilities of such equipment vary and are dependent on the hardware, including objective magnification, eyepieces, camera resolution and the software. The microscope most often encountered is a compound microscope, where the lenses rotate above the sample and light shines underneath. To effectively view samples using this instrument, samples should be flattened, most commonly onto a slide covered by a coverslip as this microscope can only focus within a single plane at a time, with anything farther or closer being out of focus.

For the purposes of imaging cyanobacteria, which are characterized by much smaller cells compared to microalgae, I have used an inverted microscope, Leica DMi 8 with LasX software, which can capture small cell species as shown in Figure 2.8. The drawback of using this type of analysis is that it relies on a destructive method, where cells are scraped off or plucked from a sample for imaging. It is also a time-consuming way of analyzing performance, as it relies on manual counting of available cells, which also greatly limits the number of samples that can be analyzed, while additionally increasing the risk of human error.

It is worth noting that mycelium, due to its typically white color, oftentimes presents a challenge when viewed under a microscope. Therefore, various stains and mounting media can be applied to aid imaging of fungal hyphae. For example, water, potassium hydroxide (KOH), lactic acid, Lacto-Fuchsin, lacto-cotton blue, modified lacto-cotton blue (MLCB), and lactophenol blue (Li, 2013). Mycelial samples have to be rehydrated if stored dry to avoid distortion, whereas cells that are already in suspension, such as microalgae liquid culture, can be placed directly onto the slide, but may require dilution in order to view individual cells.

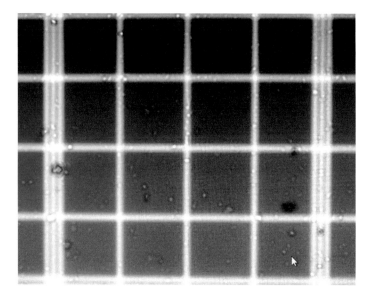

Figure 2.8 Image of cyanobacteria viewed using Leica DMi8 microscope with LasX software, each square showing 0.25 mm² white dots, indicating cyanobacterial cells in a diluted culture of 1:20, suspension culture to deionized water (D.I.) water.

Surface Imaging

Scanning electron microscopy (SEM) benefits from high depth of field and has a magnification ranging between ×10 and ×1,000,000 (Alves *et al.*, 2013). It can be used to image material surfaces, fungal spores and hyphal structures, and it captures black-and-white 3D images. SEM equipment is often found in various science and medical departments within higher education institutions and, as such, is a pertinent option when characterizing materials, as seen in Figure 2.9.

Spectroscopy

Fourier transform infrared (FTIR) spectroscopy captures the infrared spectrum of absorption or emission of a substance in various physical states. FTIR can be used to identify the chemical spectra of solids, liquids and gasses. It is used to identify materials and capture changes to chemical bonds due to breakdown of the substrate as well as identifying lipid, carbohydrate and protein content within

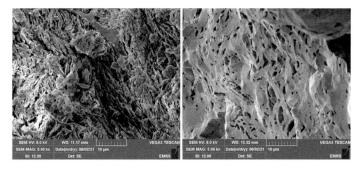

Figure 2.9 SEM images of (left) stoneware ceramic fired at 1,000°C (right) stoneware ceramic fired at 1,200°C. The difference in the structure is clearly visible, helping to identify changes in the ceramic's density caused by the firing temperature.

samples. It has been used to identify algae strains (Meng *et al.*, 2014) and fungal genus (Naumann, 2015) as well as assessing the level of decomposition or contamination of various substrates, and it is particularly useful when assessing bacterial cellulose samples.

Imaging Photosynthetic Organisms

Photosynthetic organisms, such as algae, present a challenge in terms of assessing their performance. Although individual cells can be captured using a microscope, their performance, density within a given area and changes over time are challenging to identify through purely visual means. Unlike fungi that visibly grow within days, algae may gradually change color depending on the density of the biofilm, yet it gives little insight as to whether we are observing an accumulation of dead cells or if most cells are thriving.

This is where chlorophyll florescence analysis becomes a pertinent option through pulse amplitude modulation (PAM) imaging (IMAGING-PAM M-Series; Walz GmbH, Germany), as it quantifies the level of chlorophyll fluorescence using intense light pulses to stimulate photosystem II (Schreiber, 2004). In other words, light of various wavelengths pulses over the sample, causing the photosynthetic cells to react, which, in turn, is captured by the apparatus. It is a nondistractive technique that involves placing the sample under the imaging lens and exposing it to pulsating light, from light-emitting diodes in its incubation container, without the need to alter the sample. This method offers quantitative data,

Design in the Lab

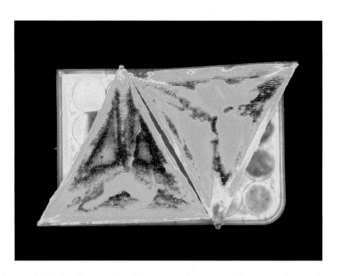

Figure 2.10 Small ceramic bio-composite pyramids, coated in algae-laden hydrogel, imaged using Imaging Pulse Amplitude Modulation (PAM) Fluorometry (I-PAM) [work by author].

alongside visual outputs, that illustrates the level of chlorophyll fluorescence in various parts of the sample, as shown in Figure 2.10. This combination is particularly useful from a design perspective as it gives a clear early indication of the development of studied bio-composites. The instrument can image photosynthetic organisms from cyanobacteria and microalgae to plants depending on the lenses available. This technique is particularly well suited to comparative studies of living cell density within a biofilm or cell migration, that is, movement of cells along surfaces over time.

BIBLIOGRAPHY

ACDP (2019) *Management and Operation of Microbiological Containment Laboratories*. London.

Alves, E. *et al.* (2013) 'Scanning electron microscopy for fungal sample examination', in *Laboratory Protocols in Fungal Biology*. New York: Springer New York, pp. 133–150. Available at: https://doi.org/10.1007/978-1-4614-2356-0-8.

Bell, S. (2009) 'Experimental design', in R. Kitchin and N. Thrift (eds) *International Encyclopedia of Human Geography*. Oxford: Elsevier, pp. 672–675. Available at: https://doi.org/https://doi.org/10.1016/B978-008044910-4.00431-4.

CDC (2021) *Infographic: Biosafety Lab Levels*, *Center for Preparedness and Response*. Available at: https://www.cdc.gov/cpr/infographics/biosafety.htm (Accessed: 19 December 2022).

Dade-Robertson, M. (2020) *Living Construction*. London: Routledge.

EngineersEdge (2022) *Fluid Volumetric Flow Rate Equation*, *Engineers Edge*. Available at: https://www.engineersedge.com/fluid_flow/volumeetric_flow_rate.htm (Accessed: 24 December 2022).

Fishel, S.R. (2017) *The Microbial State: Global Thriving and the Body Politic*. Minneapolis: University of Minnesota Press.

Gamborg, O.L. and Phillips, G.C. (1995) 'Sterile techniques', in *Plant Cell, Tissue and Organ Culture*. Berlin and Heidelberg: Springer, pp. 35–42. Available at: https://doi.org/10.1007/978-3-642-79048-5-3.

Ingold, T. (2000) *The Perception of the Environment: Essays on Livelihood, Dwelling and Skill*. London: Routledge.

Kosel, J. and Ropret, P. (2021) 'The suitability of a glovebox and of a covered still air box design for semi-sterile applications in environmental monitoring', *Journal of Microbiological Methods*, 190, p. 106325. Available at: https://doi.org/10.1016/J.MIMET.2021.106325.

Li, D.-W. (2013) 'Microscopic methods for analytical studies of fungi', in *Laboratory Protocols in Fungal Biology*. New York, NY: Springer New York, pp. 113–131. Available at: https://doi.org/10.1007/978-1-4614-2356-0-7.

Meng, Y. *et al.* (2014) 'Application of Fourier transform infrared (FT-IR) spectroscopy in determination of microalgal compositions', *Bioresource Technology*, 151, pp. 347–354. Available at: https://doi.org/10.1016/j.biortech.2013.10.064.

Meyer, M.M. (1986) 'Semiportable laminar flow hood for tissue culture and microscope use for research and teaching', *HortScience*, 21(4), pp. 1064–1065. Available at: https://doi.org/10.21273/HORTSCI.21.4.1064.

Naumann, A. (2015) 'Fourier Transform Infrared (FTIR) microscopy and imaging of fungi', in T.E.S. Dahms and K.J. Czymmek (eds). Cham: Springer International Publishing, pp. 61–88. Available at: https://doi.org/10.1007/978-3-319-22437-4-4.

Ramirez Figueroa, P.C. (2018) *Bio-Material Probes: Design Engagements with Living Systems*. University of Newcastle upon Tyne.

Sanders, Erin R. (2012a) 'Aseptic laboratory techniques: plating methods', *Journal of Visualized Experiments: JoVE*, (63), p. e3064. Available at: https://doi.org/10.3791/3064.

Sanders, Erin R. (2012b) 'Aseptic laboratory techniques: volume transfers with serological pipettes and micropipettors', *Journal of Visualized Experiments* [Preprint], (63). Available at: https://doi.org/10.3791/2754.

Sawa, M. (2016) 'The laboratory life of a designer at the intersection with algal biotechnology', *Architectural Research Quarterly*, 20(1), pp. 65–72. Available at: https://doi.org/10.1017/S1359135516000191.

Schreiber, U. (2004) 'Pulse-Amplitude-Modulation (PAM) fluorometry and saturation pulse method: an overview', in *Chlorophyll a Fluorescence*. Dordrecht: Springer Netherlands, pp. 279–319. Available at: https://doi.org/10.1007/978-1-4020-3218-9-11.

Shields, T. (2022) *Keeping It Clean: How to Design and Build a Laminar Flow Hood*, *FreshCap*. Available at: https://learn.freshcap.com/growing/keeping-it-clean-how-to-design-and-build-a-laminar-flow-hood/ (Accessed: 24 December 2022).

Staat, R.H. and Beakley, J.W. (1968) 'Evaluation of laminar flow microbiological safety cabinets', *Applied Microbiology*, 16(10), pp. 1478–1482. Available at: https://doi.org/10.1128/AEM.16.10.1478-1482.1968.

Stefanova, A. *et al.* (2020) 'Architectural laboratory practice for the development of clay and ceramic-based photosynthetic biocomposites', *Technology|Architecture + Design*, 4(2), pp. 200–210. Available at: https://doi.org/10.1080/24751448.2020.1804764.

Stefanova, A. (2021a) 'Practices in bio-design: design research through interdisciplinary collaboration', in A. Chakrabarti *et al.* (eds) *ICoRD 2021: Design for Tomorrow–Volume 3*. 1st edn. Singapore: Springer Nature Singapore Pte Ltd, pp. 41–52. Available at: https://doi.org/10.1007/978-981-16-0084-5-4.

Stefanova, A. (2021b) 'Towards a post-anthropocene bio-design practice', in A. Sharag-Eldin and C. Jarrett (eds) *ARCC 2021: Performative Environments*. Tucson: Architectural Research Centers Consortium, Inc., pp. 291-.

Working Methods

Methods

Chapter 3

DOI: 10.4324/9781003363774-3

This chapter captures basic working methods with fungus, algae and cellulose producing bacteria and is intended as a beginner's guide to initial work with microorganisms. A lot of the described methods can be hybridized or may be partially applicable to other types of lab work not covered in this volume.

WORKING WITH FUNGUS

Fungal Species: Sourcing and Considerations

There are four primary avenues for sourcing the living material or fungal culture: purchasing a commercial mycelium making product, sourcing spawn or cultures from farmers, sourcing laboratory-grade certified cultures from fungal banks or foraging samples from the wild. Each one of these supply sources comes with certain benefits and presents drawbacks that may render them unsuitable for certain design projects.

Commercial suppliers of living materials, designed for mycelium fabrication, primarily in the form of casting, offer living materials that are optimized for quick mold colonization and are often more resilient in terms of contamination. Currently, sources of such materials are limited and may only be available in certain parts of the world. They ensures predictable results but also limit possible design outcomes and variation. Presently, living materials sourced from companies such as Ecovative can be grown within five to seven days and come as a standard product, normally consisting of large particle size timber shavings that are unsuitable for certain applications such as 3D printing. To mitigate the high costs of the material, it can be mixed with fine aspen shavings at a ratio of 1:4 (Figure 3.1).

The second option is looking for mushroom farms that supply colonized living spawn or liquid culture. These sources offer a variety of liquid culture syringes that can be used to create your own supply of liquid culture or spawn. The drawback of such suppliers is that the natural species are less effective at reliably colonizing substrates and can take considerably longer to grow, compared to commercial living material. The species that can be sourced from mushroom farms are often limited to gourmet and medicinal mushroom types and are not regarded as a reliable source when conducting scientific experiments. Laboratory samples, often preserved over a long period of time and cultivated within a standardized media, cease to behave in the same responsive and efficient way as samples freshly sourced or cultivated on varied

Figure 3.1 Graduate architecture students casting using Ecovative living materials.

nutrient sources. This stagnation of growth is commonly referred to as "senescence" (McCoy, 2016). There is a growing body of research emerging within the context of bio-design, specifically utilizing such unconventional sources precisely for their responsive behavior.

The somewhat limited spectrum of fungal species available from farmers can be supplemented with foraged samples from nature. The found samples are representative of local biodiversity and have adapted to the settings within which the research is conducted. The difficulty that is often encountered within a laboratory setting is the tendency for such foraged specimens to be mycorrhizal species, reliant on symbiotic partnerships with other organisms that therefore may struggle to survive as monocultures in a lab. The specimens are also likely to remain anonymous, unless sequenced within a lab, which may require a good scientific network or may carry an associated cost.

Lastly, there are the scientific fungal collections that often sell cultures and that accept specimen deposits from researchers who may have found fungal samples in the wild. Some collections include Belgian Co-ordinated Collections of Micro-organisms (BCCM, 2022), International Collection of Microorganisms from Plants (ICMP, 2022), American Type Culture Collection (ATCC, 2022), Japan Collection of Microorganisms (JCM, 2022), National Collection of Industrial Microorganisms (NCIM, 2022) and Leibniz-Institut DSMZ-Deutsche Sammlung von Mikroorganismen und Zellkulturen GmbH (DSMZ,

2022) among others. The collections contain a much larger range of specimens that have been isolated and preserved, ensuring the reliability of the samples. Sourcing cultures from a collection is particularly appropriate for controlled laboratory testing, where the work is likely to be published within academic journals that would necessitate a repeatable setup. However as previously mentioned, cultures are often costly and need to be stored within a specific environment upon receipt, such as $-80°C$ refrigerator.

Liquid Culture

Liquid culture refers to a fungal specimen grown within a liquid nutrient broth. Fungal culture can often be purchased in liquid form and subcultured into larger containers for a continuous supply, or the culture can be started using a piece of mycelium, sourced from either an agar-grown sample or a fruiting body. Depending on the species, various types of nutrient recipes may be employed, and as with all the other forms of cultivation mentioned, certain species may not fare well in monoculture cultivation. Liquid culture is, however, a great way to cultivate mycelium for design purposes and carries lesser risks of contamination, if used within a closed system. The equipment necessary for this type of cultivation includes a glass container able to withstand high heat and pressure, a magnetic stirrer and a magnetic stirring plate, a lid with an air filter and an injection port or a lid with a hole that can be sealed using a piece of microporous tape or an air filter sticker (Figure 3.2). To transfer the culture, you will also need syringes (3–20 ml are suitable) and large gauge syringe needles (16–21 g).

There are various recipes that have been tested yielding various degrees of success. Altering the recipe for future transfers could help prevent deterioration of the culture as the fungus has to adjust to its new environment, keeping it active. A recipe I often use within my lab practice is 0.8 g light malt extract, 1 g yeast, 2 g calcium sulfate dihydrate (gypsum) per 500 ml of deionized water. To prevent contamination and reduce the need for opening the container, a magnetic stirrer and a stirring plate to stir mixtures homogeneously and to break up the mycelium mass that forms into a consolidated cluster over time should be used, allowing for the culture to be extracted using a needle and injected into spawn or onto agar.

Loosely tightened jars should be autoclaved for 60–90 minutes within an autoclave or a pressure cooker. The media should be left to

Working Methods

Figure 3.2 (Left) Mushroom growing lid with air filter and self-healing injection port. (Right) Lid with 0.6 mm hole covered by microporous tape.

cool; the lid should be securely tightened, and 0.1 ml liquid culture should be injected via a self-healing injection port within aseptic conditions. If injecting through a hole covered by microporous tape, place an additional layer of microporous tape to seal the hole left by the needle. In the event of transferring mycelium grown on agar, take a small piece of mycelium (<1 cm^2) along with the agar media, if necessary, using sanitized tweezers or a scalpel. Then lift the lid at one end enough to insert the specimen and close tightly. This should be the only time that the jar is opened while the media is grown and used. It is advisable to seal the edge of the lid with parafilm shown in Figure 3.3.

The culture should be incubated in a dark space at a temperature compatible with your selected species. Stir using a magnetic stirring plate every other day to break down lumps and incubate for two to three weeks depending on the observed level of growth. The culture should be clear with an area of white particles within a gel-like mass consolidating prior to agitation. This may be visible near the top or at the bottom portion of the jar. However, an evenly cloudy solution indicates presence of contaminants, most likely yeast. Upon reaching the desired liquid culture density, store in a refrigerator for up to six months.

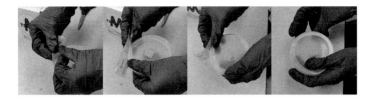

Figure 3.3 Parafilm placement in steps: (1) small unstretched piece, shorter than the area that needs to be covered; (2) grip one end of the parafilm with both hands and pull slowly about 1 cm apart distance; (3) clamp the stretched end with one thumb onto the jar and begin to rotate slowly while pulling with your other hand; (4) make sure the full perimeter is covered and some of the film is overlapping.

Agar Petri Dishes and Slants

Agar, a powder obtained from red algae is used to form a jelly consistency mass that offers a growing surface that can be used for various fundamental studies or for mycelium growth. The agar must be enriched with nutrients depending on the type of organism cultivated. The nutrient mixture and concentration will have a high impact on growth rate and organismal health.

The agar can be poured into either petri dishes or various glass or single-use plastic tubes to create slants. In both instances, the procedure must be performed within a sterile environment as contact with ambient air is likely to contaminate the agar. Petri dishes are a popular choice for mycelium cultivation as they offer a large surface area and a shallow cross section for the mycelium to grow and can be used as the first step in cultivation prior to transferring to grain or other agar plates. The petri dishes eventually dry out and therefore must be subcultured within four to six weeks. On the other hand, slants offer a deep cross section rich in nutrients and can be used to sustain samples for up to six months in most cases.

To prepare your petri dishes or slants, add 5 g light malt extract, 5 g agar and 1 g yeast to 500 ml of Deionized Water (D.I.) water and place a magnetic stirrer within the solution. Make sure you use a container that can withstand sterilization, for example, a glass beaker. Stir the mixture prior to autoclaving by using a magnetic stirring plate, and cover the opening of the container with aluminum foil. Autoclave for 15 min or more depending on quantity, at 121°C at 15 psi. or pasteurize for 2.5 hours (McCoy, 2016).

Open a new pack of petri dishes and do not open the lid until a sterile zone has been established. Pour the solution into petri dishes by slightly lifting the foil away from the opening and lifting the petri dish lid on one side, just enough to pour the liquid agar. Fill the petri dishes ½ or 2/3 of the way. Keep the petri dishes sill until the agar cools down and solidifies, forming a jellylike substance. At this point, you may observe condensation forming on the lid, obscuring the view. To remove the condensation, create stacks of 10–15 petri dishes, and place a jar of hot water or a freshly autoclaved liquid media on top of each stack (see Figure 3.4) and leave for an hour or cover with an electrical heat mat. This will remove the condensation from the lid and create a visible surface by reducing the temperature difference between the hot agar and the cooler air.

In the case of slants, make sure you use sterile tubes, either autoclaved glass tubes or single-use 50 ml falcon tubes. Fill each tube ½–2/3 of the way up, and place the tubes at a slant, so that they form a larger slanted surface upon cooling, hence the name "slants".

Once the dishes or slants are prepared, use an aseptic technique to transfer spawn, liquid culture or other samples onto the surface as described in "Aseptic Laboratory Techniques: Plating Methods" (Sanders, 2012). Once you've successfully transferred the living organism, use parafilm to seal the plates. Don't forget to label your plates with species and dates.

Tip: When subculturing from a colonized petri dish, the best areas to use for subculturing are at the edges of the mycelial matt as the

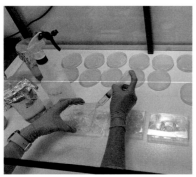

Figure 3.4 Liquid culture transfer using 30 gauge needle and syringe.

mycelium grows radially from the initial area of inoculation, leaving behind dead matter as it colonizes the surface in search of nutrients. Therefore, it is preferable to subculture from the plate before it is fully colonized.

Grain Spawn Preparation

Agar-grown mycelium or liquid culture can be used to inoculate grain spawn. This is an in-between step that allows the mycelium to propagate and take hold of the grain that will later be used with a desired substrate. There are various types of grain that can be used to cultivate mycelium; some of the most common ones include rye, corn, bird seeds and brown rice (Stamets, 2005). This is a stage that typically follows the petri dish or liquid culture phase and allows the mycelium to take hold, providing necessary nutrients for growth. If injected or transferred straight onto the bulk substrate, the mycelium may struggle to take hold; therefore, this intermediary stage is an important part of the growth process.

Grain preparation typically follows several steps to ensure that grain is sufficiently hydrated and sterile and that it contains the necessary nutrients. The grains need to be washed using a large sieve and running water over the grain for five minutes or by using a bucket and filling it with water and straining the grain multiple times. The container should then be filled with water so that it covers the grain. The following nutrients can be added to the water: 10 g gypsum, 10 g malt extract and 5 g yeast for every 4 L of water. The grain should be left for 12–24 hours, prior to draining the mixture (Figure 3.5).

A glass jar or a mushroom bag with an air filter and an injection port can be used to autoclave the grain. Let the contents fully cool down and utilize an aseptic technique to either inject 1 ml liquid culture or place a piece of colonized agar in the sterilized grain. Incubate for five to seven days prior to shaking the jar to speed up colonization and continue growing until the full mass is populated.

Note! The grain should not be too hard or too soft, and individual grains should not be bursting from their outer skin or clumping into a wet mass.

Bulk Substrate Preparation

Mushrooms usually utilize lignin, cellulose and hemicellulose as their primary food source (Gnanam, 2019). Depending on the fungal species utilized, there may be preferred substrates for

Figure 3.5 Grain being washed prior to autoclaving.

propagation that are widely used. Although many will grow on various types of agricultural wastes, such as rice husks or coco coir, or common waste sources, such as coffee grounds; not all species grow successfully on these substrates. Therefore, it is important to cross-reference the available literature to establish the appropriate type of substrate for the species you are using within your research. It is also worth noting that within design settings, there are also sources of substrate that may be abundant but less hospitable to the mycelium you are trying to propagate. For example, educational design institutions often have a supply of sawdust from their workshop facilities, which is often made up of various types of timber that has been pretreated with antifungals, toxic chemicals and adhesives that may slow down or discourage colonization altogether.

Different types of substrate material are going to influence the material properties of the product. These may include compressive strength, texture and acoustic properties, to name a few. Particle size may also play a role in the type of fabrication possible; for example, 3D printing matrices may need a more viscous and homogenous matrix than casting of large monolithic volumes.

For example, *Ganoderma lucidum* (*G. lucidum*), a species widely utilized in various design projects, typically grows on hardwoods;

therefore, untreated aspen timber shaving can be used as the bulk substrate and enriched. To promote better propagation and higher density, the substrate may be enriched with rye flower, gypsum, malt extract, yeast, potato dextrose and magnesium sulfate. A literature survey conducted by Soh *et al.* (Soh *et al.*, 2021) provides a list of common substrates and potential supplements for G. *lucidum* and *Pleurotus ostreatus* and offers a list of widely available options for cultivation.

The substrate should be sterilized or pasteurized to prevent contamination. This may involve placing the substrate in a mushroom growing bag that has a filter and autoclaving or boiling for 2–3 hours or by using ozone to remove contaminants, placing the substrate in a plastic bag along with an ozone generator and running it for 30 min. Once prepared, the substrate can be mixed with the spawn at a ratio of 4:1–6:1 w/w.

WORKING WITH MICROALGAE AND CYANOBACTERIA

Microalgae are either eukaryotic, single-cell organisms or prokaryotic simple multicellular organisms (Reynolds, 2006). They can fix between 10 and 50 times more carbon dioxide (CO_2) from the air than terrestrial plants, depending on the species (Fazal *et al.*, 2018), with 1 m^3 of algae able to match the photosynthetic efficiency of 80–100 trees (Shared Innovation, 2020). The primary challenges with microalgae cultivation, for the purposes of CO_2 fixation as well as biomass production and wastewater treatment, are the area, the amount of water and the cumbersome cultivation equipment that is often difficult to integrate within urban settings and buildings.

Cultivation Modes

There are four modes of microalgae cultivation: autotrophic, heterotrophic, mixotrophic and photoheterotrophic. Microalgae species such as *Chlorella vulgaris* can be cultivated autotrophically, and as such, they process carbohydrates through photosynthesis using CO_2 when light is available (Sajadian, Morowvat and Ghasemi, 2018). When grown in suspension, high-cell density can limit the amount of light available, resulting in a decreased rate of photosynthesis. On the other hand, heterotrophic cultivation utilizes organic carbon, where the sugar source is converted into lipids. Organisms that are exclusively heterotrophic are unable to utilize CO_2 from the air or light; however, they are often utilized within industry for the purposes of biodiesel production (Venkata Mohan *et al.*,

Working Methods

2015). Mixotrophic cultivation employs a combination of autotrophic and heterotrophic conditions for their growth. Species such as *C. vulgaris* increase their biomass yield in the presence of an organic carbon source. This mode of cultivation allows the use of lower light levels, offering the opportunity for energy saving within an artificial system (Fazal *et al.*, 2018). Lastly, in photoheterotrophic cultivation, sugar and light are utilized simultaneously, unlike mixotrophic cultivation, where either light or sugar is employed during different times (Chen *et al.*, 2011).

For the purposes of carbon dioxide sequestration, the autotrophic mode of cultivation is of particular interest, as it does not require an additional carbon source. However, studies employ mixotrophic and photoheterotrophic modes of cultivation in the context of biomass production for fuel or food.

Microalgae Suspension Cultivation Systems

Microalgae are typically cultivated within liquid cultures, that is, in suspension. These systems can be broken down into four main groups: open pond, closed pond, hybrid systems and minimal moisture environments. Open pond systems are typically open to the air tanks, 0.3–0.6 m deep and 0.8–1 m in diameter. They utilize available light to process nutrients, often from wastewater. These systems are often lower in cost; however, they are more vulnerable to contamination (Chen *et al.*, 2011). Closed pond systems, in contrast, are controlled systems where the level of nutrients, the amount of light and CO_2 are regulated. This system is more cumbersome, but it is less prone to contamination, and more predictable results can be achieved (Månsson, 2012). Hybrid systems employ both methods of suspension cultivation during different stages of the culture's development. Typically, the culture would be grown within an open pond system initially, and it would be transferred into a closed system for optimized performance at a later stage (Umar, 2018).

In the above examples, the liquid culture requires large amounts of space and large volumes of water. The integration of such systems is particularly challenging in the context of the built environment; hence, the majority of examples of algae cultivation occur within industrial contexts (Månsson, 2012; Warra, 2016; Fazal *et al.*, 2018). This highlights the benefit of cultivation of microalgae, within minimal moisture environments, which utilize higher cell density and smaller quantities of moisture, within a slurry or a gel form as discussed in Chapter 4 (In-na *et al.*, 2020; Stefanova *et al.*, 2020, 2021; Caldwell *ct al.*, 2021).

Sourcing Microalgae Culture

Cultures can be sourced from the Australian National Algae Culture Collection (CSIRO, 2022), the Culture Collection of Algae at the University of Texas in Austin (UTEX, 2022), Roscoff Culture Collection (RCC) from France (RCC, 2022) and Chlamydomonas Resource Center at the University of Minnesota (Chlamydomonas Resource Center, 2022), for example. For a full list of biological culture collections, refer to the World Federation for Culture Collections (WFCC, 2022). As with fungi, phytoplankton species can also be sourced from commercial suppliers that target hobbyists and students, and they may offer a more economical alternative, but consistency or purity is not guaranteed.

Common microalgae species used in design include *Chlorella*, *Spirulina*, *Dunaliella* and *Haematococcus* as well as bioluminescent dinoflagellates such as *Pyrosystis* (Mobin and Alam, 2017).

Cultivation and Incubation

Many facilities will have environmental controlled spaces where light levels, temperature and humidity are controlled for various types of algae strains. Although this is not necessary, it is advised to have an incubator or a setting where the light cycle and temperature can be controlled (Lourenço, 2020). It is also helpful not to cultivate algae and other organisms in the same space to reduce risks of contamination.

Photosynthetic organisms needing access to light must be cultivated in clear containers such as clear glassware, petri dishes, well plates, tubes, carboys and bottles. As the media is be sterilized prior to subculturing, it is advisable to use containers that are autoclaved or sterile.

Media Preparation

The nutrient media will vary based on the species, which can be established through a literature survey. The following example outlines cultivation of *C. vulgaris* in Blue-Green medium (BG11). To aid the process, a concentrated set of stock solutions are prepared and diluted as needed to form the nutrient media (Table 3.1).

Once the media is prepared, it should be autoclaved, preferably in the container that is going to be used for cultivation. Any tubing and tools you will use for subculturing, that is, transfer of living culture into the fresh media, should be autoclaved. The autoclaved media should cool down completely prior to subculturing. A 1:4 ratio of living culture to liquid media should be employed, for example, 1 L living culture to 4 L autoclaved liquid media. For solid media, follow

Table 3.1 BG11 recipe for 500 ml concentrated stock and subsequent amounts to be diluted per 1 L of water based on the Culture Collection of Algae and Protozoa (CCAP) recipe (CCAP, 2019)

Stock no.	Ingredient	Amount per 500 ml deionized water (Stock)	Stock per 1 L D.I. water
1	$NaNO_3$	N/A	12 g
2	K_2HPO_4	20 g	1 ml
3	$MgSO_4 \cdot 7H_2O$	37.5 g	1 ml
4	$CaCl_2 \cdot 2H_2O$	18.0 g	1 ml
5	Citric acid	3.0 g	1 ml
6	Ammonium ferric citrate green	3.0 g	1 ml
7	$EDTANa_2$	0.5 g	1 ml
8	Na_2CO_3	10 g	1 ml
9	Trace metal solution:		1 ml
	H_3BO_3	2.86 g	
	$MnCl_2 \cdot 4H_2O$	1.81 g	
	$ZnSO_4 \cdot 7H_2O$	0.22 g	
	$Na_2MoO_4 \cdot 2H_2O$	0.39 g	
	$CuSO_4 \cdot 5H_2O$	0.08 g	
	$Co(NO_3)_2 \cdot 6H_2O$	0.05 g	
Solid media			
	Agar	N/A	10 g

the previously mentioned instructions regarding plating and petri dish preparation (Figure 3.6).

To transfer the culture without opening the containers, follow Figure 3.7.

If an optimized nutrient cultivation media is not required, algae and cyanobacteria can often be cultivated using carbon dioxide tablets or sodium nitrate. This simplified nutrient broth is suitable for art installations and applications where algae performance is not being assessed.

Nutrient Sources

A wide range of nutrient options are available for the cultivation of microalgae, offering economical alternatives and opportunities for bioremediation.

Synthetic Media

Synthetic media is developed to match the nutrient requirements of various phytoplankton species. For example, there are several artificial growth media used for cultivation of *C. vulgaris*, including Hoagland's medium, Bold's Basal medium, Acidified Bold's Basal medium, Half strength Chu-10 medium and BG11 medium (Ilavarasi *et al.*, 2011). The benefit of using an engineered medium is that its chemical composition remains constant, equaling to more reliable results being obtained through laboratory experiments. However, engineered media such as BG11 carry a significant cost and are

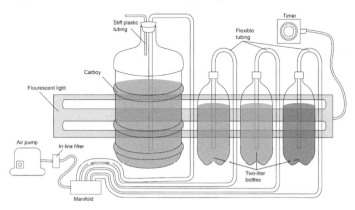

Figure 3.6 Microalgae liquid culture cultivation setup using autotrophic mode of cultivation. Diagram based on Dusjagr (2010).

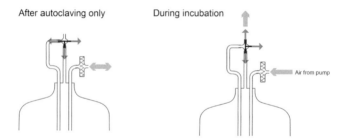

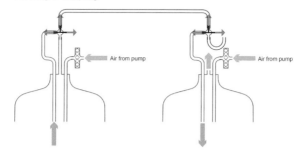

Figure 3.7 Algae culture transfer and air supply setup with a three-way stopcock.

difficult to obtain outside of a laboratory context. They are derived from pure chemical ingredients and, as such, do not provide further environmental benefits. Therefore, it is preferable to utilize compatible waste products in order to benefit from the algae's natural wastewater bioremediation abilities, hence significantly reducing costs and enabling wider-scale implementation.

Textile Wastewater as a Substitute Engineered Growth Medium
The textile industry is responsible for a large percentage of wastewater production worldwide, with countries that rely heavily on the textile industry being affected the most. An example of this is Pakistan, a country that contributes 9% to the global textile industry (Taneja, Ray and Pande, 2016), where wastewater produced by the textile industry is responsible for 30% of the country's wastewater (Murtaza and Zia, 2011).

Microalgae have been used for the purposes of wastewater remediation due to their ability to utilize CO_2, nitrogen, phosphorus and heavy metals for their metabolic functions. Textile wastewater is predominately made up of dyes, salts and heavy metals. Even though the chemical composition of the wastewater varies, depending on the type of dye, it is primarily made up of nitrogen, phosphorous and heavy metals, such as chromium (Cr), arsenic (Ar), copper (Cu) and zinc (Zn). Research shows that microalgae present an effective solution to water remediation, with a typical rate of 91% nitrogen and 93.5% phosphorous removal from nutrients, and 45% nitrogen and 32% phosphorous removal from wastewater (Zhu *et al.*, 2013).

Different species of algae have various rates of nutrient removal; therefore, the species of algae have to be selected in relationship to the type of wastewater composition to maximize results. For example, *Chlorella vulgaris* has been found to remove 44.4%–45.1% of ammonia (NH_4–N), 33.1%–33.3% of phosphate (PO_4–P) and 38.3%–62.3% of oxygen demands were reduced within 12 days. As much as 91.0% and 93.5% of nitrogen and phosphorus were removed, respectively, at peak times (Su, Mennerich and Urban, 2012). These rates are affected by the presence of toxic pollutants and variations within the chemical composition of the wastewater.

Tomato Production Waste
On average, between 800 and 1,000 m^3 of water are used to cultivate 1,000 m^2 of tomato plants within a greenhouse setting (Månsson,

2012). A study conducted by Månsson in 2012 demonstrates the potential of wastewater from tomato plants to be used for cultivation of *C. vulgaris*. It suggests the possible pairing of algae-based living surfaces with agricultural buildings or the added benefit of introducing urban agricultural practices into densely populated metropolises.

Human Urine as a Sustainable Nutrient Source for Algae Cultivation
Human urine is a possible cultivation media for microalgae such as *C. vulgaris* as it has a high phosphorous and nitrogen content. The chemical composition of human urine varies from person to person, and therefore, a synthetic alternative is typically used in laboratory experiments to ensure consistency of results. However, within an *in vivo* scenario, it provides an economical alternative, where sustainability of nutrient sources is prioritized.

Experiments have been conducted using urine that was not autoclaved or filtered, and therefore, contained bacteria. The dilution used in the study was a urine-to-water ratio of 1:100, and CO_2 was sourced from the surrounding air. The experiment was conducted in low illumination, and it demonstrates the capacity of *C. vulgaris* to grow in less-than-optimal conditions using diluted human urine (Jaatinen, Lakaniemi and Rintala, 2016). These conditions reflect an *in vivo* environment, more closely representing a plausible solution that reflects the ability of urea to sustain *C. vulgaris* in a live setting.

This approach to waste is explored through various solutions such as microbial fuel cell (MFC) toilets, that power lighting or emergency wearable technology that can produce energy using bacteria or algae (Taghavi *et al.*, 2014; Pang, Gao and Choi, 2018) or bacterial biomineralization (Arnardottir *et al.*, 2020).

Microalgae Harvesting
There are a number of extraction methods that have been commonly employed in liquid cultures. Each method has advantages and limitations, especially when employed on a large scale. Therefore, the harvesting method becomes an important factor both in separating the old cells from the substrate as well as the ability to use the harvested algae for biofuel or nutrients within the food industry (Chen *et al.*, 2011).

Sedimentation is the simplest method of harvesting. It involves collecting the sediment from the bottom of the tank. It happens over time, with an average of 0.1 and 0.2 m day^{-1} (Chen, Chang and Lee, 2015). This method is only suitable for large size cells as it relies on the gravitational force alone, making it a slow process that does not require additional energy. Similarly, algae cells can be gathered on the surface through the release of gas bubbles (primarily hydrogen or ozone) at the base of the tank. This is known as "floatation", where the cells on the surface of the solution are collected either manually or by a mechanical system (Hanotu, Bandulasena and Zimmerman, 2012). This method of harvesting has an efficiency of 50%–90% (Umamaheswari and Shanthakumar, 2016). Alternatively, liquid culture can be filtered through filters or membranes with various pore sizes, depending on the phytoplankton cell size (Rios *et al.*, 2011).

However, if a quicker process is required or if algae slurry needs to be sourced for experiments, centrifugation is the preferred method. It is a mechanical option that separates cells from liquid culture through a centrifugal force. The process can separate as much as 80%–90% within a couple of minutes (Leite and Daniel, 2020). Even though it is relatively quick and efficient in terms of cell separation, it requires an energy-powered device that has a capacity limited by its size (Mannweiler and Hoare, 1992). Flocculation, on the other hand, is a chemical method that utilizes flocculants of coagulants that are added to the algae culture. As algae tend to have a lower density than that of water, they typically float on the surface, by adding flocculants that attach to the cells, and aid sedimentation (Salim *et al.*, 2011). Electrolytic harvesting is another option, where ions are generated within the cell suspension culture; the ions help separate the cells through flocculation, relying on the negative charge of the algae cells (Misra *et al.*, 2014).

Lastly, there are various physical methods that can be employed where algae are dried using a number of drying methods, including sun drying, which is energy efficient, but may require a large surface area. Other drying methods include drum drying and freeze drying. In the case of minimal moisture environments, drying is a viable option as the cell concentration is much higher than in suspension. The cells can be scraped off the surface, providing viable area for repeat growth and harvesting (Chen, Chang and Lee, 2015).

WORKING WITH BACTERIAL CELLULOSE
Cellulose is the most abundant organic polymer present in the cell walls of most terrestrial and aquatic plants as well as fungi

(Muthu and Rathinamoorthy, 2021; Antunes *et al.*, 2022). It is found in particularly high concentrations within cotton, timber and hemp, but it is also secreted by bacteria as a metabolic by-product during fermentation (Jonas and Farah, 1998). Bacterial cellulose differs from its plant-based counterparts in its higher chemical purity and higher water absorption, and it offers a more sustainable alternative to traditional sources (Illa *et al.*, 2021). Its unique properties have come to the forefront in material research with applications emerging in biomedical settings, packaging and fashion. Bacterial cellulose films often appear translucent and range from brittle to leather-like in texture, depending on drying methods and posttreatment. These nonporous sheets are often grown on sugary media and can grow as large as the surface of the growth container. Within design fields, these new textiles present new frontiers in renewable and compostable materials, which can replace parts of the building fabric that have shorter lifespans, such as interior finishes, temporary external cladding or upholstering.

Bacterial cellulose forms as a film on the surface of a growth container. The primary factors affecting the growth of bacterial cellulose are bacterial strain, nutrient media composition, temperature, fermentation time, oxygen levels, the area of media surface, pH and light levels (Muthu and Rathinamoorthy, 2021). Other aspects that influence the properties and growth of the cellulose are incubator or bioreactor design, media agitation such as stirring or shaking and posttreatment that may include drying methods, dyes and coatings.

Strains

There is a growing body of work investigating various bacterial cellulose-producing species and strains of cellulose-producing bacteria. One of the most abundantly documented strains is *Acetobacter xylinum* (Naritomi *et al.*, 1998). Other extensively studied genera include *Sarcina* (Canale-Parola and Wolfe, 1964), *Agrobacterium* (Matthysse, 1983), *Rhizobium* (Ahmed, Kazim and Hassan, 2017), *Aerobacter*, *Azotobacter* and *Komagataeibacter* (Lin *et al.*, 2020; Aditya *et al.*, 2022; Kaczmarek, Jędrzejczak-Krzepkowska and Ludwicka, 2022). Isolated strains can be purchased from the previously mentioned microorganism banks (see section "Fungal Species: Sourcing and Considerations"). There are also emerging online vendors that cater to educators and hobbyists who supply isolated strains at a much lower price that may be more suitable for bio-design practice, but who may not offer the same level of quality control as established scientific banks. To that effect, there is also the possibility of experimenting with kombucha snobby, which is gaining popularity within scientific fields as well as design fields. The kombucha starter culture can be purchased from specialist food shops. The bacterial cellulose in the kombucha mix is produced by a mix of bacterial strains that can be sequenced to determine the exact strains. A study conducted by Kaashyap *et al.* into the composition of such a starter culture indicated that the most abundant genera in the mix were *Acetobacter*, *Bacillus* and *Komagataeibacter*, where the total included 34 genera and 200 species (Kaashyap, Cohen and Mantri, 2021). These types of cultures are widely accessible and can serve to produce bulk scoby for design purposes. The bacterial cellulose may be less pure, of a darker color or uneven in thickness in certain instances and may be less smooth, therefore requiring posttreatment to address aesthetics and material properties.

pH

One of the primary factors influencing bacterial cellulose rate of growth and health is pH. Existing studies indicate that optimal pH levels are 4–6 (Masaoka, Ohe and Sakota, 1993), although this range may vary depending on the strain used. Bacterial cellulose production sharply decreases in pH lower than 3.5 (Embuscado, Marks and BeMiller, 1994) and higher than 7 (Dirisu *et al.*, 2017), with growth cessation in pH of 3 (Muthu and Rathinamoorthy, 2021).

It is worth noting that certain biological reactions can gradually lower the pH of the media during the fermentation process. Therefore, it is recommended to monitor pH level throughout the fermentation period and to adjust with a base solution accordingly or to begin with a solution, which is within the higher end of the favorable spectrum.

Static vs. Agitated Culture and Oxygenation

The most common method of cultivating bacterial cellulose is in static culture, which is not shaken or mixed during fermentation. This results in a bacterial cellulose sheet that gradually forms on the surface of the culture, where the liquid and air come into contact (Hsieh *et al.*, 2016). In this mode of cultivation, the oxygen content limits the rate of growth of the pellicle (Hornung *et al.*, 2006).

Agitating the culture, on the other hand, can result in other types of bacterial cellulose formation. For example, continuously shaking the culture can produce individual balls of cellulose or bespoke bioreactors at times designed to continuously mix the culture. Further consideration should be given to the method of oxygenation of the culture as it can result in bacterial cellulose forming around air outlets, or increasing the oxygen content of the culture as it can also increase yield, as evidenced by various studies (Hwang *et al.*, 1999; Chao *et al.*, 2000) (Figure 3.8).

Temperature
Temperature is a major factor in bacterial cellulose production which affects both the viability of culture and the rate of growth. Viability according to studies ranges from 20°C to 40°C, with higher temperatures preventing bacterial cellulose production (Rangaswamy, Vanitha and Hungund, 2015). Optimal temperature ranges vary based on the bacterial strain but fit within the wider range of 25°C–30°C, with significant decreases in the rate of growth in temperatures above 30°C (Son *et al.*, 2001; Zhao, Li and Zhu, 2018).

Fermentation Time
The necessary time for bacterial cellulose cultivation varies depending on nutrient type and amount, strain and incubation conditions. Static kombucha culture has a fermentation time that ranges between 18 and 21 days, after which bacterial production is either halted or significantly reduced due to nutrients and oxygen depletion of the culture (Goh *et al.*, 2012). However, various studies indicate that the yield of bacterial cellulose and fermentation time

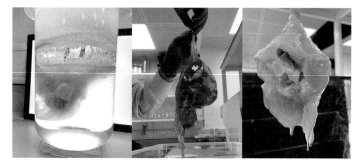

Figure 3.8 Bacterial cellulose forming within the culture as a result of placing air supply outlets within the culture. Work and image by Aileen Hoenerloh.

differ based on incubation conditions and nutrient composition (Keshk, 2014; Costa and Santos, 2017; Yanti, Ahmad and Muhiddin, 2018; de Amorim *et al.*, 2020).

Nutrient Media

Bacterial cellulose requires carbon and nitrogen sources that can be sourced from various tea types (Amarasekara, Wang and Grady, 2020), coffee types (Fontana *et al.*, 1991), fruit juices (Calderón-Toledo *et al.*, 2022; Muhajir *et al.*, 2022), wastewater from industries such as rice water, tofu water (Srikandace *et al.*, 2022), molasses (Esin Poyrazoğlu Çoban, 2011) and other sucrose-, glucose- and fructose-based sources. In addition, various additives can be used to increase production yields, including glycerol, ethanol, acetic acid and lactic acid (Agustin and Padmawijaya, 2018).

Hestrin and Schramm (HS) medium is the standard synthetic media developed in the 1950s for the cultivation of bacterial cellulose (Hestrin and Schramm, 1954). The following recipe is for 1 L of deionized water: 20 g d-glucose, 5 g peptone, 5 g yeast extract, 1.15 g citrate and 2.7 g disodium phosphate (15 g agar if making solid medium for petri dishes). The media should be adjusted using sodium hydroxide and a pH meter until a pH of 6.0 is achieved. The media should then be autoclaved prior to use. This type of media is more prone to contamination but produces purer bacterial cellulose pellicles.

A commonly used alternative to HS media is an enriched tea broth, like that used for kombucha brewing. A recipe I often use in my own work includes 2 g tea (black or green), 100 g granulated sugar, 75 ml apple cider vinegar and 90 ml glycerol per 1 L boiling deionized water. A small piece (2 cm^2) of the starter scoby is added with 50 ml of the starter broth to the cooled media. It is worth noting that kombucha scoby is comparatively easy to cultivate, although it is prone to contamination in multispecies environments, for example, if mycelium work is taking place; therefore, the higher vinegar content in the above recipe aims to deter potential contamination. If the space is not used for work with other microorganisms, the vinegar content can be decreased to 25 ml/1 L (vinegar/water).

Whether using a synthetic culture or alternatives from available sources, you will need to establish a refeeding cycle as the nutrients are depleted. This may involve adding fresh media on top of the formed sheet to create a continuous culture, where a new sheet will begin to form or media can be fed underneath a growing sheet using plastic tubing and an intravenous drip (IV) bag.

Harvesting, Drying and Posttreatments

Harvesting the bacterial cellulose pellicle entails washing the sheet to remove any excess media. This can be done using soap or various acidic or alkaline solutions (Tang *et al.*, 2010). A protocol for purifying the fully grown pellicle, as described by Muthu and Rathinamoorthy, involves harvesting the pellicle and submerging it in a 1%–6% sodium hydroxide solution for a minimum of an hour or up to two days, depending on the thickness of the pellicle. This is then followed by an acetic acid bath lasting 15 min to 12 hours to neutralize the base solution. Finally, the sheet is washed with tap water to remove excess chemicals (Muthu and Rathinamoorthy, 2021).

Following a purification process, the pellicle may be further treated to alter its color by bleaching using hydrogen peroxide (Suryanto *et al.*, 2018) or sodium peroxide, of 5% concentrations. These treatments influence the tensile properties of the material with sodium peroxide, increasing the elasticity of the material. This is also a good stage to introduce dyes of natural or artificial origin, through submerging the pellicle in a bath containing the desired agent. Various studies have also been conducted to compare this posttreatment approach to dyeing versus dyeing during fermentation, which often results in uneven coloration and may affect the growth conditions.

The next step in the process is drying the pellicle, which can be done in several ways that may further impact material properties. Drying methods may involve air drying on porous or nonporous material; for example, if drying on top of a piece of cardboard, the sheet is going to dry as a brittle, smoother layer that must be removed by rewetting to separate it from the base. This contrasts with air drying on top of a plastic sheet, which would result in a slower drying process that offers greater flexibility within the sheet and an uneven surface, due to uneven heat distribution.

Drying can be followed by a plasticizing treatment to further alter the material properties, namely, to reduce rehydration of the sheet. This can be done by submerging the sheet in a 10% glycerol bath over a 24-hour period to give the sheet greater flexibility. Once dried, the sheet should be heat pressed or ironed to lock in the glycerol and give the sheet a more uniform look. Ensure you protect the bacterial cellulose with a Teflon sheet from the surface of the heated element. The glycerol acts as a plasticizer in this case, which also prevents water reabsorption upon wetting. This glycerol plasticizing treatment can also be applied to pure mycelium sheets. Other plasticizers include polyvinyl alcohol (Sun *et al.*, 2018), citric acid and epichlorohydrin (Almeida *et al.*, 2022).

Apart from evaporation, there are various other methods including oven drying, freeze drying and microwaving, to name a few. The drying method can have an impact on material properties and will be suitable for different applications and methods of analysis and assessment. From an experimental standpoint, oven drying provides a quantifiable drying time and temperature. Freeze drying is a more cumbersome process, but the advantage is that the sample retains its shape to a greater degree. Therefore, if the sample is grown through a process that affects the morphology, the shape can be recorded more effectively through freeze drying. Microwaving the sample is like evaporation; however, the drying time is significantly reduced to a matter of minutes, and the sample assumes a more uniform, smoother texture due to even heat distribution.

BIBLIOGRAPHY

Aditya, T. *et al.* (2022) 'Surface modification of bacterial cellulose for biomedical applications', *International Journal of Molecular Sciences*, 23(2), p. 610. Available at: https://doi.org/10.3390/ijms23020610.

Agustin, Y.E. and Padmawijaya, K.S. (2018) 'Effect of acetic acid and ethanol as additives on bacterial cellulose production by acetobacter xylinum', *IOP Conference Series: Earth and Environmental Science*, 209, p. 012045. Available at: https://doi.org/10.1088/1755-1315/209/1/012045.

Ahmed, S.A., Kazim, A.R. and Hassan, H.M. (2017) 'Increasing cellulose production from rhizobium leguminosarum bv. viciae', *Journal of Al-Nahrain University-Science*, 20(3), pp. 120–125. Available at: https://doi.org/10.22401/JNUS.20.1.17.

Almeida, A.P.C. *et al.* (2022) 'Crosslinked bacterial cellulose hydrogels for biomedical applications', *European Polymer Journal*, 177, p. 111438. Available at: https://doi.org/10.1016/j.eurpolymj.2022.111438.

Amarasekara, A.S., Wang, D. and Grady, T.L. (2020) 'A comparison of kombucha SCOBY bacterial cellulose purification methods', *SN Applied Sciences*, 2(2), p. 240. Available at: https://doi.org/10.1007/s42452-020-1982-2.

de Amorim, J.D.P. *et al.* (2020) 'Plant and bacterial nanocellulose: production, properties and applications in medicine, food, cosmetics, electronics and engineering. A review', *Environmental Chemistry Letters*, 18(3), pp. 851–869. Available at: https://doi.org/10.1007/s10311-020-00989-9.

Antunes, F.A.F. *et al.* (2022) 'The potential of vegetal biomass for biomolecules production', in *Comprehensive Renewable Energy*. Elsevier, pp. 139–164. Available at: https://doi.org/10.1016/B978-0-12-819727-1.00053-4.

Arnardottir, T. *et al.* (2020) *Turbulent Casting Bacterial Expression in Mineralized Structures*. ACADIA.

ATCC (2022) *American Type Culture Collection, ATCC*. Available at: https://www.atcc.org/ (Accessed: 27 September 2022).

BCCM (2022) *Belgian Coordinated Collections of Microorganisms, BCCM*. Available at: https://bccm.belspo.be/ (Accessed: 27 September 2022).

Calderón-Toledo, S. *et al.* (2022) 'Isolation and partial characterization of Komagataeibacter sp. SU12 and optimization of bacterial cellulose production using Mangifera indica extracts', *Journal of Chemical Technology & Biotechnology*, 97(6), pp. 1482–1493. Available at: https://doi.org/10.1002/jctb.6839.

Caldwell, G.S. *et al.* (2021) 'Immobilising microalgae and cyanobacteria as biocomposites: new opportunities to intensify algae biotechnology and bioprocessing', *Energies*, 14(9), p. 2566. Available at: https://doi.org/10.3390/en14092566.

Canale-Parola, E. and Wolfe, R.S. (1964) 'Synthesis of cellulose by Sarcina ventriculi', *Biochimica et Biophysica Acta (BBA) – General Subjects*, 82(2), pp. 403–405. Available at: https://doi.org/10.1016/0304-4165(64)90314-9.

CCAP (2019) *BG11 (Blue-Green Medium), Culture Collection of Algae and Protozoa*. Available at: www.ccap.ac.uk/wp-content/uploads/MR_BG11.pdf.

Chao, Y. *et al.* (2000) 'Bacterial cellulose production by Acetobacter xylinum in a 50-L internal-loop airlift reactor', *Biotechnology and Bioengineering*, 68(3), pp. 345–352. Available at: https://doi.org/10.1002/(SICI)1097-0290(20000505)68:3<345::AID-BIT13>3.0.CO;2-M.

Chen, C.-L., Chang, J.-S. and Lee, D.-J. (2015) 'Dewatering and drying methods for microalgae', *Drying Technology*, 33(4), pp. 443–454. Available at: https://doi.org/10.1080/07373937.2014.997881.

Chen, C.-Y. *et al.* (2011) 'Cultivation, photobioreactor design and harvesting of microalgae for biodiesel production: a critical review', *Bioresource Technology*, 102(1), pp. 71–81. Available at: https://doi.org/10.1016/J.BIORTECH.2010.06.159.

Chlamydomonas Resource Center (2022) *Chlamydomonas Resource Center at the University of Minnesota, Chlamydomonas Resource*

Center. Available at: https://www.chlamycollection.org/ (Accessed: 27 December 2022).

Costa, R. and Santos, L. (2017) 'Delivery systems for cosmetics – from manufacturing to the skin of natural antioxidants', *Powder Technology*, 322, pp. 402–416. Available at: https://doi.org/10.1016/j.powtec.2017.07.086.

CSIRO (2022) *Australian National Algae Culture Collection, CSIRO*. CSIRO.

Dirisu, C. *et al.* (2017) 'pH effect and pH changes during biocellulose production by gluconacetobacter xylinus in moringa oleifera tea-sugar medium', *Journal of Advances in Microbiology*, 7(2), pp. 1–7. Available at: https://doi.org/10.9734/JAMB/2017/38440.

DSMZ (2022) *German Collection of Microorganisms and Cell Cultures GmbH, Leibniz Institute*. Available at: https://www.dsmz.de/ (Accessed: 27 September 2022).

Dusjagr (2010) *Plankton, Hackteria Wiki*. Available at: https://hackteria.org/wiki/File:Plankton.jpg (Accessed: 29 December 2022).

Embuscado, M.E., Marks, J.S. and BeMiller, J.N. (1994) 'Bacterial cellulose. II. Optimization of cellulose production by Acetobacter xylinum through response surface methodology', *Food Hydrocolloids*, 8(5), pp. 419–430. Available at: https://doi.org/10.1016/S0268-005X(09)80085-4.

Esin Poyrazoğlu Çoban (2011) 'Evaluation of different pH and temperatures for bacterial cellulose production in HS (Hestrin-Scharmm) medium and beet molasses medium', *African Journal of Microbiology Research*, 5(9). Available at: https://doi.org/10.5897/AJMR11.008.

Fazal, T. *et al.* (2018) 'Bioremediation of textile wastewater and successive biodiesel production using microalgae', *Renewable and Sustainable Energy Reviews*, 82, pp. 3107–3126. Available at: https://doi.org/10.1016/j.rser.2017.10.029.

Fontana, J.D. *et al.* (1991) 'Nature of plant stimulators in the production of Acetobacter xylinum ("tea fungus") biofilm used in skin therapy', *Applied Biochemistry and Biotechnology*, 28–29(1), pp. 341–351. Available at: https://doi.org/10.1007/BF02922613.

Gnanam, C. (2019) *Introduction to Mycology*. 1st edition. Triplicane: MJP Publishers.

Goh, W.N. *et al.* (2012) 'Fermentation of black tea broth (Kombucha): I. Effects of sucrose concentration and fermentation time on the yield of microbial cellulose', *International Food Research Journal*, 19, pp. 109–117.

Hanotu, J., Bandulasena, H.C.H. and Zimmerman, W.B. (2012) 'Microflotation performance for algal separation', *Biotechnology and Bioengineering*, 109(7), pp. 1663–1673. Available at: https://doi.org/10.1002/bit.24449.

Hestrin, S. and Schramm, M. (1954) 'Synthesis of cellulose by Acetobacter xylinum. 2. Preparation of freeze-dried cells capable of polymerizing glucose to cellulose', *Biochemical Journal*, 58(2), pp. 345–352. Available at: https://doi.org/10.1042/bj0580345.

Hornung, M. *et al.* (2006) 'Optimizing the production of bacterial cellulose in surface culture: evaluation of substrate mass transfer influences on the bioreaction (Part 1)', *Engineering in Life Sciences*, 6(6), pp. 537–545. Available at: https://doi.org/10.1002/elsc.200620162.

Hsieh, J.-T. *et al.* (2016) 'A novel static cultivation of bacterial cellulose production by intermittent feeding strategy', *Journal of the Taiwan Institute of Chemical Engineers*, 63, pp. 46–51. Available at: https://doi.org/10.1016/j.jtice.2016.03.020.

Hwang, J.W. *et al.* (1999) 'Effects of pH and dissolved oxygen on cellulose production by Acetobacter xylinum BRC5 in agitated culture', *Journal of Bioscience and Bioengineering*, 88(2), pp. 183–188. Available at: https://doi.org/10.1016/S1389-1723(99)80199-6.

ICMP (2022) *International Collection of Microorganisms from Plants, Manaaki Whenua*. Available at: https://www.landcareresearch.co.nz/tools-and-resources/collections/icmp-culture-collection/ (Accessed: 27 September 2022).

Ilavarasi, A. *et al.* (2011) 'Optimization of various growth media to freshwater microalgae for biomass production', *Biotechnology*, 10, pp. 540–545. Available at: https://doi.org/10.3923/biotech.2011.540.545.

Illa, M.P. *et al.* (2021) 'In situ tunability of bacteria derived hierarchical nanocellulose: current status and opportunities', *Cellulose*, 28(16), pp. 10077–10097. Available at: https://doi.org/10.1007/s10570-021-04180-3.

In-na, P. *et al.* (2020) 'Loofah-based microalgae and cyanobacteria biocomposites for intensifying carbon dioxide capture', *Journal of CO2 Utilization*, 42, p. 101348. Available at: https://doi.org/10.1016/j.jcou.2020.101348.

Jaatinen, S., Lakaniemi, A.-M. and Rintala, J. (2016) 'Use of diluted urine for cultivation of Chlorella vulgaris', *Environmental Technology*, 37(9), pp. 1159–1170. Available at: https://doi.org/10.1080/09593330.2015.1105300.

JCM (2022) *Japan Collection of Microorganisms, RIKEN BRC*. Available at: https://jcm.brc.riken.jp/en/aboutjcm_e (Accessed: 27 September 2022).

Jonas, R. and Farah, L.F. (1998) 'Production and application of microbial cellulose', *Polymer Degradation and Stability*, 59(1–3), pp. 101–106. Available at: https://doi.org/10.1016/S0141-3910(97)00197-3.

Kaashyap, M., Cohen, M. and Mantri, N. (2021) 'Microbial diversity and characteristics of kombucha as revealed by metagenomic and physicochemical analysis', *Nutrients*, 13(12), p. 4446. Available at: https://doi.org/10.3390/nu13124446.

Kaczmarek, M., Jędrzejczak-Krzepkowska, M. and Ludwicka, K. (2022) 'Comparative analysis of bacterial cellulose membranes synthesized by chosen Komagataeibacter strains and their application potential', *International Journal of Molecular Sciences*, 23(6), p. 3391. Available at: https://doi.org/10.3390/ijms23063391.

Keshk, S.M.A.S. (2014) 'Vitamin C enhances bacterial cellulose production in Gluconacetobacter xylinus', *Carbohydrate Polymers*, 99, pp. 98–100. Available at: https://doi.org/10.1016/j.carbpol.2013.08.060.

Leite, L. de S. and Daniel, L.A. (2020) 'Optimization of microalgae harvesting by sedimentation induced by high pH', *Water Science and Technology*, 82(6), pp. 1227–1236. Available at: https://doi.org/10.2166/wst.2020.106.

Lin, D. *et al.* (2020) 'Bacterial cellulose in food industry: current research and future prospects', *International Journal of Biological Macromolecules*, 158, pp. 1007–1019. Available at: https://doi.org/10.1016/j.ijbiomac.2020.04.230.

Lourenço, S. de O. (2020) 'Microalgae culture collections, strain maintenance, and propagation', in *Handbook of Microalgae-Based Processes and Products*. Elsevier, pp. 49–84. Available at: https://doi.org/10.1016/B978-0-12-818536-0.00003-8.

Mannweiler, K. and Hoare, M. (1992) 'The scale-down of an industrial disc stack centrifuge', *Bioprocess Engineering*, 8(1–2), pp. 19–25. Available at: https://doi.org/10.1007/BF00369259.

Månsson, S. (2012) *Cultivation of Chlorella Vulgaris in Nutrient Solution from Greenhouse Tomato Production*. Swedish University of Agricultural Sciences.

Masaoka, S., Ohe, T. and Sakota, N. (1993) 'Production of cellulose from glucose by Acetobacter xylinum', *Journal of Fermentation and Bioengineering*, 75(1), pp. 18–22. Available at: https://doi.org/10.1016/0922-338X(93)90171-4.

Matthysse, A.G. (1983) 'Role of bacterial cellulose fibrils in Agrobacterium tumefaciens infection', *Journal of Bacteriology*, 154(2), pp. 906–915. Available at: https://doi.org/10.1128/jb.154.2.906-915.1983.

McCoy, P. (2016) *Radical Mycology: A Treatise on Seeing and Working with Fungi*. Portland: Chthaeus Press.

Misra, R. *et al.* (2014) 'Electrochemical harvesting process for microalgae by using nonsacrificial carbon electrode: a sustainable approach for biodiesel production', *Chemical Engineering Journal*, 255, pp. 327–333. Available at: https://doi.org/10.1016/j.cej.2014.06.010.

Mobin, S. and Alam, F. (2017) 'Some promising microalgal species for commercial applications: a review', *Energy Procedia*, 110, pp. 510–517. Available at: https://doi.org/10.1016/J.EGYPRO.2017.03.177.

Muhajir, M. *et al.* (2022) 'Effect of homogenization pressure on bacterial cellulose membrane characteristic made from pineapple peel waste', *Journal of Mechanical Engineering Science and Technology (JMEST)*, 6(1), p. 34. Available at: https://doi.org/10.17977/um016v6i12022p034.

Murtaza, G. and Zia, M.H. (2011) *Wastewater Production, Treatment and Use in Pakistan*. Faisalabad.

Muthu, S.S. and Rathinamoorthy, R. (2021) 'Bacterial cellulose', pp. 19–60. Available at: https://doi.org/10.1007/978-981-15-9581-3-2.

Naritomi, T. *et al.* (1998) 'Effect of lactate on bacterial cellulose production from fructose in continuous culture', *Journal of Fermentation and Bioengineering*, 85(1), pp. 89–95. Available at: https://doi.org/10.1016/S0922-338X(97)80360-1.

NCIM (2022) *National Collection of Industrial Microorganisms, National Chemical Laboratory*. Available at: https://www.ncl-india.org/files/NCIM/Default.aspx (Accessed: 27 September 2022).

Pang, S., Gao, Y. and Choi, S. (2018) 'Flexible and stretchable biobatteries: monolithic integration of membrane-free microbial fuel cells in a single textile layer', *Advanced Energy Materials*, 8(7), p. 1702261.

Rangaswamy, B.E., Vanitha, K.P. and Hungund, B.S. (2015) 'Microbial cellulose production from bacteria isolated from rotten fruit', *International Journal of Polymer Science*, 2015, pp. 1–8. Available at: https://doi.org/10.1155/2015/280784.

RCC (2022) *Roscoff Culture Collection, RCC*. Available at: https://www.roscoff-culture-collection.org/about-rcc (Accessed: 27 December 2022).

Reynolds, C. (2006) *Ecology of Phytoplankton*. 1st edn. Cambridge: Cambridge University Press.

Rios, S.D. *et al.* (2011) 'Dynamic microfiltration in microalgae harvesting for biodiesel production', *Industrial & Engineering Chemistry Research*, 50(4), pp. 2455–2460. Available at: https://doi.org/10.1021/ie101070q.

Sajadian, S., Morowvat, M. and Ghasemi, Y. (2018) 'Investigation of autotrophic, heterotrophic, and mixotrophic modes of cultivation on lipid and biomass production in Chlorella vulgaris', *National Journal of Physiology, Pharmacy and Pharmacology*, 8(5), p. 1. Available at: https://doi.org/10.5455/njppp.2018.8.0935625122017.

Salim, S. *et al.* (2011) 'Harvesting of microalgae by bio-flocculation', *Journal of Applied Phycology*, 23(5), pp. 849–855. Available at: https://doi.org/10.1007/s10811-010-9591-x.

Sanders, E.R. (2012) 'Aseptic laboratory techniques: plating methods', *Journal of Visualized Experiments: JoVE*, 63, p. e3064. Available at: https://doi.org/10.3791/3064.

Shared Innovation (2020) *Growing Algae on Façades: It Works!, Bouygues Innovation*. Available at: https://www.bouygues-construction.com/blog/en/reflexions-prospectives/algues-facade-culture/ (Accessed: 22 May 2021).

Soh, E. *et al.* (2021) 'Effect of common foods as supplements for the mycelium growth of Ganoderma lucidum and Pleurotus ostreatus on solid substrates', *PLOS ONE*, 16(11), p. e0260170. Available at: https://doi.org/10.1371/journal.pone.0260170.

Son, H.-J. *et al.* (2001) 'Optimization of fermentation conditions for the production of bacterial cellulose by a newly isolated Acetobacter sp.A9 in shaking cultures', *Biotechnology and Applied Biochemistry*, 33(1), p. 1. Available at: https://doi.org/10.1042/BA20000065.

Srikandace, Y. *et al.* (2022) 'Bacterial cellulose production by Komagataeibacter xylinus using rice-washed water and tofu processing wastewater with the addition of sodium glutamate', *Fibers and Polymers*, 23(5), pp. 1190–1196. Available at: https://doi.org/10.1007/s12221-022-4729-4.

Stamets, P. (2005) *Mycelium Running: How Mushrooms Can Help Save the World*. New York: Ten Speed Press.

Stefanova, A. *et al.* (2020) 'Architectural Laboratory practice for the development of clay and ceramic-based photosynthetic biocomposites', *Technology|Architecture + Design*, 4(2), pp. 200–210. Available at: https://doi.org/10.1080/24751448.2020.1804764.

Stefanova, A. *et al.* (2021) 'Photosynthetic textile biocomposites: using laboratory testing and digital fabrication to develop flexible living building materials', *Science and Engineering of Composite Materials*, 28(1), pp. 223–236. Available at: https://doi.org/10.1515/secm-2021-0023.

Su, Y., Mennerich, A. and Urban, B. (2012) 'Synergistic cooperation between wastewater-born algae and activated sludge for wastewater treatment: influence of algae and sludge inoculation ratios', *Bioresource Technology*, 105, pp. 67–73. Available at: https://doi.org/10.1016/J.BIORTECH.2011.11.113.

Sun, Y. *et al.* (2018) 'The effects of two biocompatible plasticizers on the performance of dry bacterial cellulose membrane: a comparative study', *Cellulose*, 25(10), pp. 5893–5908. Available at: https://doi.org/10.1007/s10570-018-1968-z.

Suryanto, H. *et al.* (2018) 'Effect of peroxide treatment on the structure and transparency of bacterial cellulose film', *MATEC Web of Conferences*. Edited by P. Puspitasari et al., 204, p. 05015. Available at: https://doi.org/10.1051/matecconf/201820405015.

Taghavi, M. *et al.* (2014) 'Wearable self sufficient MFC communication system powered by urine', in M. Mistry et al. (eds) *Advances in Autonomous Robotics Systems*. Cham: Springer, pp. 131–138. Available at: https://doi.org/10.1007/978-3-319-10401-0_12.

Taneja, N., Ray, S. and Pande, D. (2016) *India – Pakistan Trade: Textiles and Clothing*. New Delhi. Available at: https://doi.org/http://hdl.handle.net/10419/176353.

Tang, W. *et al.* (2010) 'The influence of fermentation conditions and post-treatment methods on porosity of bacterial cellulose membrane', *World Journal of Microbiology and Biotechnology*, 26(1), pp. 125–131. Available at: https://doi.org/10.1007/s11274-009-0151-y.

Umamaheswari, J. and Shanthakumar, S. (2016) 'Efficacy of microalgae for industrial wastewater treatment: a review on operating conditions, treatment efficiency and biomass productivity', *Reviews in Environmental Science and Bio/Technology*, 15(2), pp. 265–284. Available at: https://doi.org/10.1007/s11157-016-9397-7.

Umar, A. (2018) *The Screening, Fabrication and Production of Microalgae Biocomposites for Carbon Capture and Utilisation*. Newcastle University.

UTEX (2022) *UTEX Culture Collection of Algae*, *UTEX*. Available at: https://utex.org/ (Accessed: 27 December 2022).

Venkata Mohan, S. *et al.* (2015) 'Heterotrophic microalgae cultivation to synergize biodiesel production with waste remediation: progress and perspectives', *Bioresource Technology*, 184, pp. 169–178. Available at: https://doi.org/10.1016/J.BIORTECH.2014.10.056.

Warra, A.A. (2016) 'Degradation of environmental contaminants through bioremediation: a review', *Journal of Biotechnological Research*, 1(2), pp. 89–100. Available at: https://doi.org/10.20448/805.1.2.89.100.

WFCC (2022) *World Federation for Culture Collections*, *WFCC*. Available at: https://www.wfcc.info/ (Accessed: 27 December 2022).

Yanti, N.A., Ahmad, S.W. and Muhiddin, N.H. (2018) 'Evaluation of inoculum size and fermentation period for bacterial cellulose production from sago liquid waste', *Journal of Physics: Conference Series*, 1116, p. 052076. Available at: https://doi.org/10.1088/1742-6596/1116/5/052076.

Zhao, H., Li, J. and Zhu, K. (2018) 'Bacterial cellulose production from waste products and fermentation conditions optimization', *IOP Conference Series: Materials Science and Engineering*, 394, p. 022041. Available at: https://doi.org/10.1088/1757-899X/394/2/022041.

Zhu, L.D. *et al.* (2013) 'Recycling harvest water to cultivate Chlorella zofingiensis under nutrient limitation for biodiesel production', *Bioresource Technology*, 144, pp. 14–20. Available at: https://doi.org/10.1016/J.BIORTECH.2013.06.061.

Experiments in Design

Chapter 4

DOI: 10.4324/9781003363774-4

MYCELIUM CASTING

Mycelium casting involves the use of living material to form a shape determined by a static mold. The best way to think of this is to imagine plaster or concrete that is formed using a formwork. However, there are certain differences in the making of the formwork that pertain to the living organism. As we have already established, mycelium requires air, and thus, deep cross sections cannot be formed without considering the organism's air supply. Typically, a large volume may be formed by casting smaller portions and stacking them together, while still alive, to allow them to fuse into a single mass, a process known as "myco-welding". Another method is perforating the mold to maximize airflow; however, this will not ensure deep cross-section colonization. For that to occur, an air supply would have to be introduced into the deeper regions of the mold. This may be done by designing hollow sections and creating perforated internal walls or employing more experimental methods, such as inserting perforated plastic tubing in between the substrate, which can be attached to an air pump to circulate oxygen into the deeper parts of the mold.

Incubation of the mycelium casts is also a crucial part of the process, where temperature, humidity and storage have a profound effect on both the speed of colonization and the texture of the cast. In most instances, the mold is wrapped in cling film or covered with a plastic bag, which must be perforated in multiple places to ensure access to air. The incubation setting may be a growth tent or a designated enclosed space, which is dark as well as temperature and humidity controlled. Humidity within the chamber should be kept at 60%–80%, and a heater without open heating elements can be placed within the tent to maintain favorable conditions, which, in the case of *Ganoderma lucidum*, is a range of 24°C–29°C.

Mold making is another aspect of the design that must be considered in terms of material compatibility. Mycelium being wet and living, needs to remain moist during the incubation process. It grows particularly well within nonporous structures such as vacuum formed plastic sheets, prefabricated plastic containers and PLA 3D printed vessels. However, this is predicated on the use of plastics and thus may not meet sustainability objectives. Porous materials are at times used to varying degrees of success where materials such as untreated timber, paper and card, to name a few, offer economical alternatives, but may present certain pitfalls. Porous materials are prone to contamination during incubation as they are difficult to disinfect prior to filling with living material, and they can also draw moisture out of the substrate. A particular difficulty that may present itself is removing the material from the mold, if the mold material is not flexible, and the design does not take this into consideration. Other material alternatives include textiles that can be grown in tension, using gravity and acting in compression once dried as shown in Figure 4.5.

Another example is sheet casting where the material is grown flat as a layer on top of a supporting surface, which, following a five-to-seven-day period, is pulled to form a curved geometry. This is followed by further incubation and drying, while maintaining the same curvature as shown in Figures 4.1 and 4.2.

A typical process of casting involves creating a mold that is sprayed with alcohol prior to infilling. The living material that has colonized the substrate is then broken up into small pieces and filled within the mold. The openings of the mold should then be covered with cling film, which is then perforated multiple times, or a lid that has an air filter or micro holes. The cast is then incubated in a dark warm environment, with a humidity of 60%–80%. Incubation

Figure 4.1 Mycelium cast using a vacuum formed plastic sheet. Work by Studio V M.Arch students, University of Colorado Denver, work done under the direction of the Author.

Figure 4.2 (Left to Right) Curved mycelium cast using plastic packaging, Y- and X-shaped structural members cast in 3D printed formwork. Work by Studio V M.Arch students, University of Colorado Denver, work done under the direction of the Author.

Figure 4.3 Prototype of timber frame infilled with mycelium insulation. Work by Studio V M.Arch students, University of Colorado Denver, work done under the direction of the Author.

time may vary in the rate of myceliation, which is determined by the species used and the substrate type. Once full myceliation is achieved, the mold should be taken out of the incubation environment and can be partially air dried, while still inside the mold, by removing any coverings, or it can be taken out of the mold immediately. Drying can occur in several ways, either by air drying,

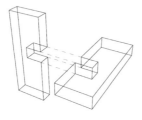
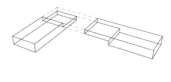

Edge Lap Joint - used to connect frame members within each module

End Lap Joint - used to connect modules to adjacent modules

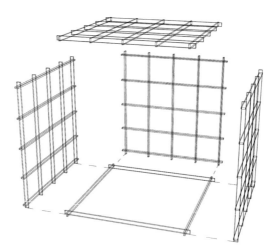

Variability in Form - modules are planned to be designed as a cube with 5 faces, but can be altered within the constraints of a modular framing system

Figure 4.4 Design of timber frame structure infilled with mycelium insulation. Work by Studio V M.Arch students, University of Colorado Denver, work done under the direction of the Author.

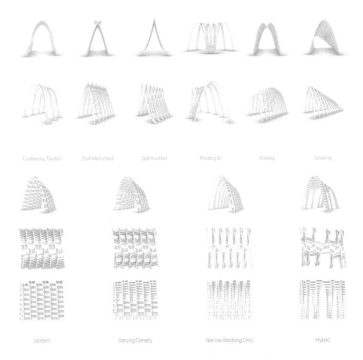

Figure 4.5 Arch design explorations based on mycelium casting using textiles. Work by Studio V M.Arch students, University of Colorado Denver, work done under the direction of the Author.

by oven drying or using a hairdryer in case of a humid climate. It is worth noting that rapid drying can, at times, deform the cast or alter its color, giving it golden and brown tints. Drying the mycelium cast over a couple of hours makes the material less reactive to the environment and more difficult to be reanimated, when in contact with high-moisture settings.

Student Project: Mycelium Making

In the Fall of 2022, I conducted a mycelium making studio with a group of 11 M.Arch students at the University of Colorado Denver. The students undertook initial material explorations to familiarize themselves with lab practice and the behavior of the living materials. This was followed by small group proposals based on the fundamental principles of mycelium casting, which resulted in two

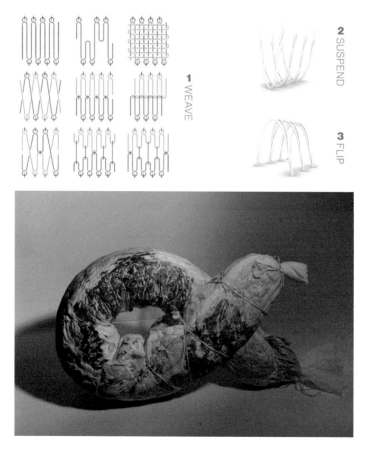

Figure 4.6 (Top) Arch design explorations based on mycelium casting using textiles, casting sequence diagram. (Bottom) Prototype of loop cast using mycelium and textile. Work by Studio V M.Arch students, University of Colorado Denver, work done under the direction of the Author.

proposals. The first proposal employed mycelium as an infill material cast *in situ* into a timber frame. The second explored the potential of mycelium to be cast using textiles to form a compressive element, such as an arch. The final part of the course culminated in a group design and build project that resulted in a unit-based mycelium structure entirely cast using laser cut paper molds (Figures 4.3–4.9).

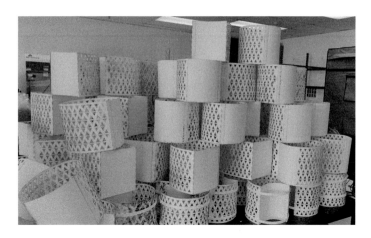

Figure 4.7 Laser cut paper molds. Work by Studio V M.Arch students, University of Colorado Denver, work done under the direction of the Author.

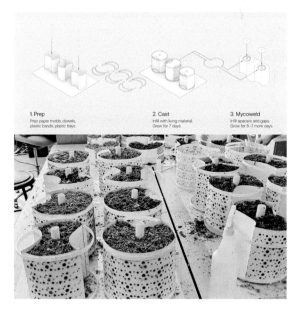

Figure 4.8 (Top) Mycelium modules fabrication plan. (Bottom) Laser cut paper molds filled with living Ecovative material prior to incubation. Work by Studio V M.Arch students, University of Colorado Denver, work done under the direction of the Author.

Experiments in Design

Figure 4.9 Mycelium structure made up of myco-welded modules. Work by Studio V M.Arch students, University of Colorado Denver, work done under the direction of the Author.

EXTRUSION-BASED 3D PRINTING

Digital fabrication with living materials presents a new frontier in making and opens the possibility to subvert established making practices as well as testing materials. Clay 3D printers utilize pressure to extrude a viscous or a paste-like consistency material. Successful extrusion is predicated on the consistency, texture and homogeneity of the matrix, and it adds a new set of constraints, in addition to those of the living organisms that need a compatible environment to colonize a substrate. The following projects illustrate different making practices that utilize 3D printing.

Clay 3D Printing for Coral Restoration

Humanity has historically adapted its habitats to suit human comfort, a process that has made human populated areas hostile environments to most other species. This has become ever more apparent in contemporary urban environments. Our global society has not only impacted terrestrial landscapes but it has also created great disturbances in the oceans. Coral bleaching events (Brown, 1997) and pollution from industrial sources, such as oil spills, (Ramseur, 2010) or plastic pollution, (Ballerini *et al.*, 2018) have brought local marine ecologies into the Anthropocene. Provided there is a lack of further interference, as proposed by researchers, is likely to result in patchy ecologies, rather than ecologies that can return to normal (Hughes *et al.*, 2017).

These emerging trends call for an active approach to the coral reefs through restoration and selective breeding of resilient colony members (Crawford, Humanes, *et al.*, 2022). This project looks at the design and manufacture of substrates for coral larvae growth using a clay 3D printer. This type of work seeks to assist natural processes and, in doing so, assists in sustaining and restoring not only the natural environment but also local communities that depend on these ecosystems.

Plug Design Requirements

The initial design stages involved conversations with the scientific team that helped define the requirements for larvae growth. The four threats that were identified during the initial consultation included grazing fish that may kill the juvenile coral prior to reaching maturity (Omori and Iwao, 2014), movement between the substrate and the rock which would prevent the coral from taking hold, algae fouling that creates biofilms forming on the surface of the substrate as well as sediment.

Geometry Design

The geometries developed measure less than 4 cm^3 and incorporate protected areas where the coral can take hold, without being affected by predatory grazing fish. The four designs that emerged focus to varying extents on the defined criteria and have different potential benefits as well as drawbacks (Figures 4.10–4.13).

Variations and On-Site Testing Considerations

To facilitate benchmarking and assessment of geometry variation, each of the four options was designed with three heights that allowed to benchmark performance against similar geometries. Each geometry varied in height by an average of ±1.5 cm, providing a controlled variable suitable for scientific studies. Although this method of standardization offered a benchmark, the testing of the substrates is split into two parts: (1) deposition of larvae within a laboratory setting and (2) on-site outplanting, following an initial cultivation period.

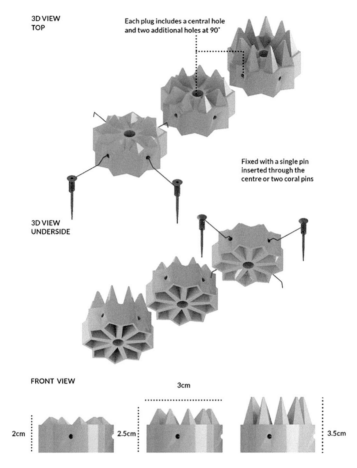

Figure 4.10 Starfish – tackles the threat of grazing fish, gives the option for larvae to grow along the star-shaped perimeter and can be affixed in two different ways: (1) with spikes facing upward and 2) with spikes facing downward.

Larvae Attachment

Once the larvae have formed, the CoralPlugs can be left under water for larvae to attach in two possible ways: (1) by placing the plugs on a tray with the desired side facing upward, allowing larvae to settle on the side to be protected, or (2) by skewering the geometries on a rod, a method that would encourage larvae settlement along the perimeter.

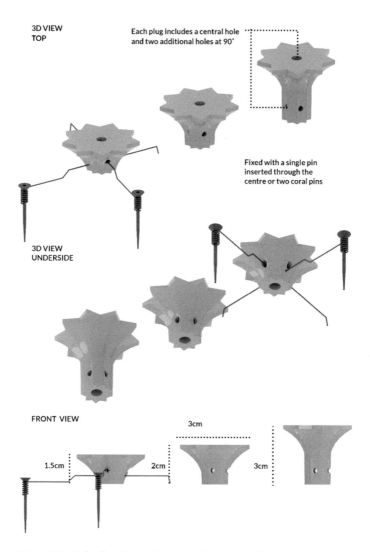

Figure 4.13 Umbrella – this option provides less shelter from grazers and presents challenges in terms of time needed for fabrication as an increased number of direction changes increase the print time.

Experiments in Design

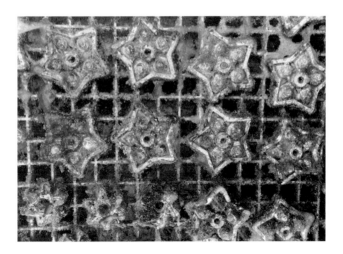

Figure 4.14 Coral larvae settlement on CoralPlugs in nursery tank.

Figure 4.15 Twisted Cog-shaped CoralPlug outplanted on the reef with juvenile coral growth. [Image by Eveline van derSteeg].

operational costs and training. The quicker and less time-consuming the procedure, the more viable and impactful the system becomes. Each design allowed two methods of attachment onto the rock: (1) using a drill and ceramics screw, secured with adhesive, and (2) a Coralclip (Suggett *et al.*, 2020), consisting of a metal wire wrapped around a nail that is fixed onto rocks using a hammer (Figures 4.15–4.17).

Figure 4.16 Mushroom CoralPlug outplanted on the reef with juvenile coral growth. [Image by Eveline van derSteeg].

Figure 4.17 Star-shaped CoralPlug outplanted on the reef with juvenile coral growth. [Image by Eveline van derSteeg].

CoralPlug Fabrication

This type of work builds upon the existing examples of human-made-made coral restoration structures deployed around the world (Jaap, 2000; Guest *et al.*, 2014; Omori and Iwao, 2014). The designs were modeled in 3D Studio Max 2020 and were imported into Slicer as an obj. file that was prepared for 3D printing with standard Lutum 4.5 3 mm presets. The clay used for fabrication was 157–1,142 White

Special Stoneware, sourced from Potclays Limited, a smooth, textured off-white color clay, which was fired at 1,200°C, using an Ecotop 43 L-UK kiln by Helmut Rohde GmbH. The wet clay was diluted with 10% water to achieve a consistency suitable for 3D printing. To fabricate the prototypes, an air pressure-based extrusion 3D printer (Lutum 4.5; VormVrij) was used with a pressure setting of 50 psi. To optimize the printing time, the large 3 mm nozzle was used, which resulted in a decreased level of detail. Slight changes were applied to the designs to optimize printing speed and to enable printing with a larger 3 mm nozzle, which is both faster and allows printing with clay that has a lower moisture content, resulting in less distortion compared to the smaller 1.2 mm nozzle. Although the designs were altered, their primary design features remained the same, and therefore, the variables that may affect coral growth, predator behavior and protection from weather conditions also remained the same (Figures 4.18 and 4.19).

Conclusion

The CoralPlugs take an active approach to restoration, where scientists and designers come together to create viable solutions that tackle the effects of climate change. The project has helped demonstrate how architects can employ their skills, within the creation of nonhuman habitats and proposes an expanded definition of the role of the architect, in the context of combating climate

Figure 4.18 (Left) 1.2 mm nozzle printing tests of umbrella design. (Right) 3 mm print in progress of starfish design.

Figure 4.19 (Left) Starfish and mushroom design 3D printed CoralPlugs. (Right) Twisted Cog design 3D printed CoralPlugs.

change. It also brings up the question of the role of restoration within the Anthropocene and the ethics of introducing humanmade structures into natural environments. This type of activity is altering ecologies and natural landscapes, further enforcing a view that humanity has continuously shaped the natural environment and that there is scope for such interventions to be eco-centric and beneficial.

Bio-composites and Photosynthetic Organisms

Encapsulation of microorganisms in organic and synthetic substrates is a biomimetic principle that is present in nature. An example of such functions are diatoms, a group of single-celled algae that self-encapsulate through the production of a shell, a process known as "biomineralization" (Heveran et al., 2020). In addition to biomineralization, a wide variety of microorganisms such as fungi, cyanobacteria and algae also colonize solid surfaces temporarily or permanently to protect themselves from adverse environments (Holzmeister et al., 2018). Within modern applications, microorganisms are often combined with organic or inorganic substrates to increase their resilience and to protect them from sudden environmental changes, to maximize their performance and to reduce the amount of space they need to function (Stefanova, et al., 2019). Within the context of the built environment, the ability to attach microorganisms onto substrates opens the possibility of integrating and sustaining metabolic functions within the various layers of the building fabric. In the following two examples, algae are combined with textiles and ceramic components to create living building surfaces.

3D Printing with Bio-laden Hydrogels: Photosynthetic Textile Bio-composites

The following projects showcase two strands of investigation: digitally fabricating with a living matrix, as demonstrated in this study, and digitally fabricating with the substrate, presented in the following section of this chapter. This investigation captures the process of developing algae-laden matrices suitable for extrusion-based 3D printing onto a textile substrate. The full project was published in *Science and Engineering of Composite Materials*; for full method and results, refer to the full paper (Stefanova *et al.*, 2021). The methods used in this study were based on protocols used to develop algae-based bio-composites (In-na *et al.*, 2020b).

Digital fabrication with bio-matrices has been an area of investigation primarily within medical and scientific studies, which oftentimes take place on a small scale (Mark Co and Culaba, 2019; Poomathi *et al.*, 2020). However, 3D printing with biomaterials has also captured the imagination of designers and artists. Various methods for digital fabrication and its application in design have been explored using mycelium (Fairs, 2013), bacterial cellulose (Chiujdea and Nicholas, 2020) and algae (Morris, 2017). The following studies fit into an existing culture of matrix development and experimental 3D printing within design (Shaklova and Cruz, 2015; Ali and Majeed, 2018; Malik *et al.*, 2020). Furthermore, this project continues to develop ways for growing algae within low moisture environments (Mohammadifar *et al.*, 2018; Pang, Gao and Choi, 2018; In-na *et al.*, 2020b; Stefanova, Bridgens, Armstrong, *et al.*, 2020), which calls for the development of cell deposition systems that are effective and that lend themselves to integration within the building realm.

Substrate

Several textile substrates were selected that differed in texture, weight and absorbance. The chosen substrates included cotton, polyester, canvas and hessian. These textiles presented widely available and economically accessible options, which are capable of absorbing and distributing nutrients as well as being suitable for various interior applications, such as wall finishes, tensile structures and partitions (Table 4.1).

Table 4.1 An overview of the results from the textile characterization using a Leica DMi 8 microscope image, using LasX software, pH probe and microscales (Stefanova *et al.*, 2021)

Textile type	Leica DMi 8 microscope image using LasX software (×5 magnification)	Thread diameter (µm)	pH	Absorption (ml/1 cm²)
Cotton		213 (mean Standard Deviation (StDev) = 1.960)	7.66 (mean StDev = 0.213)	0.0429 (mean StDev = 0.0021)
Polyester		322 (mean StDev = 22.360)	8.36 (mean StDev = 0.247)	0.1269 (mean StDev = 0.0080)

(*Continued*)

Table 4.1 (Continued)

Canvas (300 gsm)		367 (mean StDev = 43.848)	7.40 (mean StDev = 0.139)	0.0930 (mean StDev = 0.0076)
Hessian		1,115 (mean StDev = 157.849)	7.32 (mean StDev = 0.062)	0.1680 (mean StDev = 0.0138)

Matrix Development and Compatibility

In the early stages, several matrices were tested based on the protected coatings developed earlier. The bio-gel matrices incorporate the use of kappa-carrageenan (a hydrocolloid polymer extracted from red seaweeds), chitosan (a polymer commonly extracted from the exoskeletons of marine crustaceans), aloe vera (extracted from the succulent plant *Aloe barbadensis*) and a clay-based paint binder (Auro 331) (Stefanova *et al.*, 2021). An initial investigation was conducted, which evaluated the matrix performance based on compatibility with *Chlorella vulgaris*, contamination and the ability to modify the consistency of the mixture for 3D printing (Figure 4.20).

Figure 4.20 Kappa carrageenan bio-gels in 3D printing canisters.

3D Printing

Following initial investigations, the two preferred options for 3D printing compatibility development were kappa-carrageenan and kappa-carrageenan with Auro clay paint. For this study, I used an air pressure extrusion 3D printer (Lutum 4.5; VormVrij) that required a minimum of 10 psi pressure for extrusion. The print tests were done using a 0.6 mm diameter nozzle, suitable for use in conjunction with low-viscosity materials. The process involved the trial of multiple viscosities to ascertain the optimal matrix composition; the results of the tests are recorded in Table 4.2.

Table 4.2 3D printing mixtures tested, extrusion conditions and matrix behavior during mechanical extrusion (Stefanova *et al.*, 2021)

Bio-gel mixture	Extrusion pressure (psi)	Nozzle size (diameter in mm)	Comments
1. Kappa-carrageenan 0.6 g/10 ml, BG11, *C. vulgaris* slurry 0.2 ml/10 ml gel	10 psi	0.6 mm	Continuous flow from the nozzle prior to printing, occurring due to pressurization of canister
2. Kappa-carrageenan 0.8 g/10 ml, BG11, *C. vulgaris* slurry 0.2 ml/10 ml gel	10 psi	0.6 mm	Steady flow upon extrusion, gel maintaining its structure without crumbling or uncontrolled distortion
3. Kappa-carrageenan 1 g/10 ml, BG11, *C. vulgaris* slurry 0.2 ml/10 ml gel	20 psi	0.8 mm	Inconsistent flow upon extrusion, crumbling of extruded filament and poor adhesion to textile surface
4. Kappa-carrageenan 0.6 g/10 ml, 50% w/w BG11, 50% w/w Auro clay paint, *C. vulgaris* slurry 0.2 ml/10 ml gel	10 psi	0.6 mm	Continuous flow from the nozzle prior to printing, occurring due to pressurization of canister

Table 4.2 (Continued)

Bio-gel mixture	Extrusion pressure (psi)	Nozzle size (diameter in mm)	Comments
5. Kappa-carrageenan 0.8 g/10 mL, 50% w/w BG11, 50% w/w Auro clay paint, *C. vulgaris* slurry 0.2 mL/10 mL gel	10 psi	0.6 mm	Mousse-like consistency, steady flow, good adhesion to textiles
6. Kappa-carrageenan 1 g/10 ml, 50% w/w BG11, 50% w/w Auro clay paint, *C. vulgaris* slurry 0.2 ml/10 ml gel	20 psi	0.8 mm	Inconsistent flow, crumbling of filament and poor adhesion

The design of the matrix involved a balancing act of creating a stable mixture that was not prone to flaking or uncontrolled flow and the ability of the mixture to sustain living *C. vulgaris* cells. The most promising mixtures were numbers 2 and 5, which were used to conduct a set of laboratory tests.

Material Performance

The successful fabrication process was used to print on four types of textiles, including cotton, hessian, polyester and linen, using three 2D designs to produce triplicate biological samples, along with triplicate control samples. The 3D printed samples were cut into circular pieces following printing and were incubated within 90 mm petri dishes, along with 5 ml of dH$_2$O at Day 0 and regular rehydration every other day using a spray bottle. The samples were incubated in a controlled environment with a closed lid to prevent sudden evaporation. Results were collected every other day using Imaging Pulse-Amplitude-Modulation (I-PAM) Fluorometry (I-PAM) (Figure 4.21).

From the data, it becomes apparent that a carrageenan only matrix can sustain a constant number of *C. vulgaris* cells, whereas the Auro binder mixture exhibited an increase in chlorophyll fluorescence after Day 4. By using I-PAM to image the textiles, I was able to capture the cell migration that occurred over the two weeks, which was not fully visible to the naked eye, but was clear within the visual data, as shown in Figure 4.22.

Figure 4.21 3D printed algae-laden hydrogels on various types of textiles incubated in petri dishes.

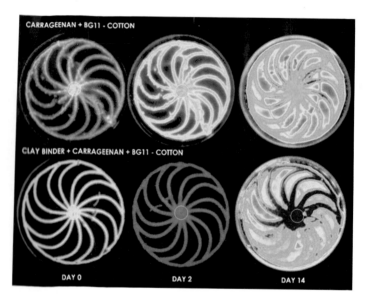

Figure 4.22 3D printed patterns on cotton using kappa-carrageenan (top) and Auro Clay Paint (bottom), image showing cell chlorophyll fluorescence in I-PAM. Red and yellow indicate low levels of fluorescence of living photosynthetic cells, lighter shades of gray indicate higher levels and black indicates a lack of living photosynthetic cells (Stefanova et al., 2021).

Storage, Transportation and Application

Although the living textiles presented a working solution within a controlled environment, there are potential design difficulties associated with transportation, storage or even the ability to place the textiles on a desired surface or frame, due to the paste-like consistency of the wet mixture, which is fragile and prone to smudging. This problem led to an investigation into the drying and subsequent rehydration of such textile-based bio-composites. Rehydration was initiated after a range of drying times (24, 48, 72 and 96 hours) and subsequent incubation of 14 days.

Based on the results from the study (Figure 4.23), there is evidence that the developed textile bio-composites could be prefabricated offsite and activated once installed within a building setting. However, during the drying process, there was a high level of flaking and damage to the printed design; therefore, further matrix development is necessary to overcome the brittle nature of the material and to reduce shrinkage during the drying process.

3D Ceramic Photosynthetic Bio-composites

The term "bio-composite" is used to refer to a combination of living organisms or their by-products or a combination of a living organism or its by-products and an inorganic or nonliving material. Bio-composite materials are providing a new pallet of products for a sustainable building fabric that is grown and that performs functions that were previously executed by mechanical systems, as in the case of CO_2 sequestration (Stefanova, Bridgens, In-na, *et al.*, 2020). The study builds on the existing research of using minimal moisture environments that sustain photosynthetic organisms through the hygroscopic properties of material substrates such as loofah, a naturally occurring plant structure (In-na *et al.*, 2020a), and various textiles (Stefanova *et al.*, 2021) (Figure 4.24).

Method
The methods utilized in this study are based on the existing protocols used to develop algae-based bio-composites (In-na *et al.*, 2020b; Stefanova, Bridgens, Armstrong, *et al.*, 2020; Stefanova *el al.*, 2021). The bio gel matrices incorporate the use of kappa-carrageenan, a hydrocolloid polymer extracted from red seaweeds and a clay-based paint binder (Auro 331). The study was split into two parts: (1) kappa carrageenan and Auro binder-based matrices on ceramic vessels with various wall thicknesses fired at 1,200°C and 2) kappa carrageenan and Auro binder-based matrices on ceramic vessels with various wall thicknesses fired at 1,000°C. The two sets of experiments were used to establish the most favorable

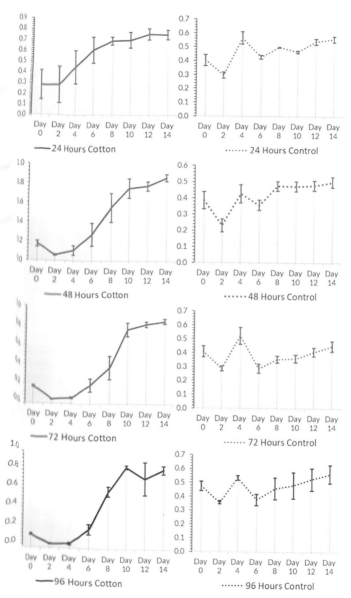

Figure 4.23 The effect of variable drying times on chlorophyll fluorescence in cotton and Auro binder bio-composites.

Experiments in Design

Figure 4.24 3D printed ceramic vessels with a high-complexity geometry, showing a loss of detail and printing inaccuracy.

substrate thickness and firing temperature. In addition to that, we compared the effects of the two types of mixtures in relation to the health of the living cells. This was achieved by assessing the effect of various types of geometries on the chlorophyll fluorescence levels of *C. vulgaris*.

Substrate Design and Preparation

In this study, the porosity of the clay is used to distribute moisture and nutrients to create a suitable environment for microalgal growth. The components presented in this project contain a hollow center that is filled with nutrients through a small, round opening at the top. For this study, 157–1,142 White Special Stoneware clay (sourced from https://www.potclays.co.uk/white-special-sware) was used, as previous experiments suggest that clays with smoother, lower aggregate makeup lend themselves better to minimal moisture cultivation of *C. vulgaris* (Stefanova, Bridgens, Armstrong, et al., 2020; Stefanova, Bridgens, In-na, et al., 2020).

pH Study and Compatibility Testing

Chlorella vulgaris flourishes in a neutral pH; therefore, it was critical to measure the pH of the selected clay to determine its compatibility. Three separate samples were tested to ensure consistency. Each sample weighed approximately 5 g and was dissolved through manual mixing in 10 ml of distilled water. The pH readings were obtained with a glass probe pH meter.

In addition, the algae culture was assessed for compatibility against various types of clay to establish an appropriate substrate type (Figures 4.25 and 4.26).

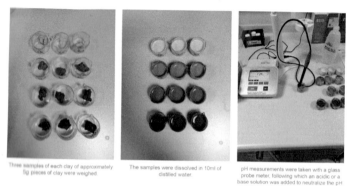

Figure 4.25 pH testing of various types of clay using a pH meter.

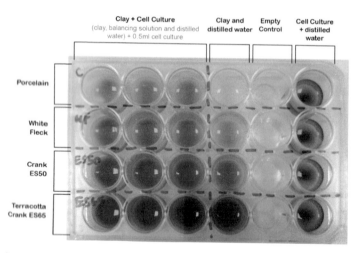

Figure 4.26 24-Well plate setup. Image shows the toxicity setup; there are three samples of each clay with cell culture, along with a clay control, empty control and a cell culture control. The samples were stirred with a pipette and photographed daily.

Experiments in Design

Geometry

Three orthographic pyramid geometries were developed for testing the health of the algae: (1) smooth, (2) inset folding and (3) outward folding. The three geometries were based on the same simple orthogonal pyramid, measuring 12 cm × 12 cm × 3 cm. The variations serve to inform our knowledge of the effect of larger surface area and the behavior of the algae, when encountering inward and outward folded structures.

Wall Thickness

In addition to the geometry of the pyramids, four different wall thicknesses were tested, including one, two, three and four wall layers, each 3 mm thick. This was done to establish the effect on water distribution in relationship to wall thickness. For each wall thickness, three geometries were tested, creating a set of 12 samples. Wall thickness was added internally, reducing the internal space but maintaining the same outer shell dimensions (Figures 4.27 and 4.28).

Firing Temperature

Ceramics have different porosity, depending on the temperature of firing in a kiln. The kiln used was the Ecotop 43 L-UK from Helmut Rohde GmbH. For these studies, two sets of vessels with three types of geometries and four wall thicknesses were generated and fired at different temperatures: (1) full firing at 1,250°C and (2) bisque firing at 1,000°C. The higher temperature results in reduced porosity and

Figure 4.27 3D printed ceramic vessel samples, and three geometries printed with four different wall thickness, one, two, three and four wall layers, each 3 mm thick. The three geometries were based on the same simple orthogonal pyramid, measuring 12 cm × 12 cm × 3 cm

Figure 4.28 3D printing of various geometry samples with different wall thicknesses.

greater strength of the ceramic. The following ceramics' firing cycles were used: (1) full firing cycle of 60°C per hour until reaching 600°C, followed by 250°C per hour until reaching 1,250°C, and a gradual cooling, upon reaching maximum temperature, lasting 14 hours; (2) bisque firing cycle of 30°C per hour until 250°C, 50°C per hour, until 600°C, followed by 250°C per hour, until reaching a maximum temperature of 1,000°C and then gradually cooling for 12 hours.

Clay Preparation and 3D Printing

The geometries were modeled in 3D Studio Max 2018 and exported as OBJ files into Slicer 3D software to generate a 3D printing file for 3 mm nozzle clay printing. The printer used was the Lutum 4.5 air-based extrusion clay printer at a pressure of 60 psi. Wet clay sold in 12 kg bags was mixed with 10% water to achieve a soft consistency compatible with the 3D printing equipment. The clay was manually loaded into 2 kg cartridges.

Algae Matrix Preparation

The inoculating liquid algae culture was grown in full-strength Blue-Green medium (BG11) (Stanier *et al.*, 1971) at 18 ± 2°C with a 16:8-hour light-to-dark photoperiod, at a light intensity of 2,500 lux (~30.5 µmol m^{-2} s^{-1}) (Thimijan and Heins, 1983) provided by 30W daylight-type fluorescent tubes (Sylvania Luxline Plus; n = 6). The cultures were placed in 50 ml Falcon tubes and centrifuged at 1,620 RCF (relative centrifugal force) for 10 minutes to produce a dense algae slurry. The algae slurry was mixed with the tested gel matrix at a ratio of 0.05 ml algae slurry per 1 ml of gel (Stefanova, Bridgens, In-na, *et al.*, 2020).

Each bio-gel matrix was prepared in 20 ml batches in 50 ml beakers. For the baseline kappa-carrageenan treatment, 0.8 g of kappa-carrageenan (from >99.9% pure powder; Sigma-Aldrich, UK) was added to 20 ml (0.04 g/ml) of full-strength BG11 and stirred at room temperature. Once a gel-like consistency was achieved, 0.4 ml of *C. vulgaris* slurry was added and stirred until homogeneous. Auro 331 Clay Paint binders (AURO Paint Company, UK) were tested as additives to the kappa-carrageenan baseline. For the Auro paint, 0.64 g of kappa-carrageenan was dissolved in 16 ml of BG11 (0.04 g/ml) across the previous dilution series, to which 10 ml of Auro 331 was added (equivalent to 50% w/w Auro content) (Stefanova *et al.*, 2021).

Setup and Incubation

A layer of the two mixtures was added using a paintbrush, covering the whole vessel evenly, leaving the base and a strip of 1 cm around the base clear. Each vessel was placed within a plastic well-plate lid. The vessels were filled with 40 ml BG11 using a syringe, and an additional 10 ml BG11 was added to the well plate.

The samples were incubated at room temperature with each sample set of ceramics and controls placed in a non-airtight, clear plastic box, measuring 47×35×24 cm and sprayed with D.I. water once every 24 hours. A 45W plant growth light with 52 red and 28 blue lamp beds was used to illuminate the samples on a 16-hour light eight hours dark cycle. Control samples were tested in triplicate as follows: BG11 and algae slurry suspension samples, algae slurry and carrageenan, algae slurry and carrageenan and Auro binder. Control samples were cultivated in six well plates that remained open for the duration of the test and were incubated under the same conditions as the ceramic samples (Figures 4.29 and 4.30).

Along with an analysis of the numerical data, the samples were visually compared across the 14-day period using the visual output from I-PAM, and photographs were taken prior to imaging.

Results

Ceramic Substrate and Firing Temperature

In this study, the samples fired at 1,000°C outperformed the samples fired at 1,200°C, although both sets of samples exhibited healthy functions for the duration of the experiment. In this set

Figure 4.29 Application of algae-laden matrices: (left) kappa-carrageenan and (right) kappa-carrageenan and Auro clay paint binder.

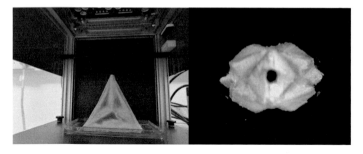

Figure 4.30 (Left) Bio-sample placed within imaging chamber of I-PAM. (Right) I-PAM image of coated ceramic samples.

of experiments, firing temperature was a significant factor, and geometry wall thickness was also significant, while the interaction between the two variables was not significant (Crawford, In-na, *et al.*, 2022). There is a clear differentiation between samples that were fired at 1,200°C and 1,000°C, the former retaining less water and prone to higher levels of evaporation due to the binder content that dries quicker than a carrageenan gel matrix.

Binder and Kappa Carrageenan Matrix
The translucent nature of the carrageenan mixture allowed imaging of cells that were locked within the matrix using PAM, which appeared as a higher density of cells in the early stage, visible to the naked eye and showing clearly in I-PAM. The Auro binder matrix offered an impermeable surface, where the color of the mixture was considerably lighter, and the cells located below the surface were not imaged. Over the 14-day period, more saturated patches began to appear in the case of 1,000°C fired ceramic samples. The distribution indicated that cells grew away from inward folds and that smooth surfaces or outward folds offered a more hospitable environment. This is likely due to higher levels of light exposure at the outward folds, along with lower level of algae cell wash-off, which typically occurs at inward folding channels.

Discussion
Although there is a level of resilience that comes from the design of hollow components that can hold a reserve of water and nutrients, surface evaporation occurs quickly and is likely to be exacerbated by air movement as well as types of inhabitation points needed for an evaporation retardation strategy. The design of a full building setup goes beyond the scope of this study and merits separate investigation in an *in vivo* experiment. However, this is likely to take the form of either a rigid or a tensile layer, which can create an air pocket that limits the amount of air movement and evaporation. Such a setup may direct and trap moisture from sources such as humidifiers or natural evaporation from water bodies, for example, interior water features.

Conclusion
The study demonstrates a method for digital manufacture of ceramic substrates for living building bio-composites. It reveals how the design of the geometry, wall thickness and firing temperature of the substrate actively influence the behavior and performance of living photosynthetic microalgae. The performance of the components is influenced predominantly by evaporation and moisture levels, and therefore, the results point to a need for a vapor control system. The experiments offer vital fundamental data that can inform further development of commercially viable, full system designs (Figure 4.31).

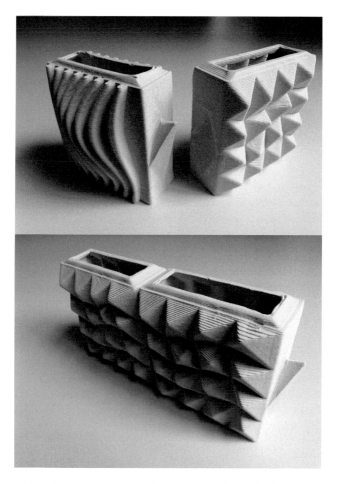

Figure 4.31 Interlocking ceramic building units demonstrating a potential wall structure based on the presented study.

BIBLIOGRAPHY

Ali, S. and Majeed, S. (2018) 'Advancement of bio inks in three dimensional bioprinting', *Biomedical Journal of Scientific & Technical Research*, 11(4), pp. 001–004. Available at: https://doi.org/10.26717/bjstr.2018.11.002129.

Ballerini, T. *et al.* (2018) *Plastic Pollution in the Ocean: What We Know and What We Don't Know About*. thecamp. Available at: https://doi.org/10.13140/RG.2.2.36720.92160.

Brown, B.E. (1997) 'Coral bleaching: causes and consequences', *Coral Reefs*, 16, pp. S129–S138. Available at: https://doi.org/10.1007/s003380050249.

Chiujdea, R.S. and Nicholas, P. (2020) 'Design and 3D printing methodologies for cellulosebased composite materials', in D. Werner, Liss C. Köring (ed.) *Anthropologic – Architecture and Fabrication in the Cognitive Age*. Berlin: eCAADe, pp. 547–558.

Crawford, A., Humanes, A., Caldwell Svan der Steeg, E., and Guest, J. (2022) 'Architecture for coral restoration: using clay-based digital fabrication to overcome bottlenecks to coral larval propagation', in Cruz, H. (ed.) *ICSA 2022, 5th International Conference on Structures and Architecture*. Aalborg, pp. 458–466. Available at: https://doi.org/10.1201/9781003023555-55.

Crawford, A., In-na, P., *et al.* (2022) 'Clay 3D printing as a bio-design research tool: development of photosynthetic living building components', *Architectural Science Review*, 65(3), pp. 185–195. Available at: https://doi.org/10.1080/00038628.2022.2058908.

Fairs, M. (2013) *Mycelium Chair by Eric Klarenbeek is 3D-printed with Living Fungus, Dezeen*. Available at: https://www.dezeen.com/2013/10/20/mycelium-chair-by-eric-klarenbeek-is-3d-printed-with-living-fungus/ (Accessed: 22 October 2020).

Guest, J.R. *et al.* (2014) 'Closing the circle: is it feasible to rehabilitate reefs with sexually propagated corals?', *Coral Reefs*, 33(1), pp. 45–55. Available at: https://doi.org/10.1007/s00338-013-1114-1.

Heveran, C.M. *et al.* (2020) 'Biomineralization and successive regeneration of engineered living building materials', *Matter*, 2(2), pp. 481–494. Available at: https://doi.org/10.1016/j.matt.2019.11.016.

Holzmeister, I. *et al.* (2018) 'Artificial inorganic biohybrids: the functional combination of microorganisms and cells with inorganic materials', *Acta Biomaterialia*, 74, pp. 17–35. Available at: https://doi.org/10.1016/j.actbio.2018.04.042.

Hughes, T.P. *et al.* (2017) 'Coral reefs in the Anthropocene', *Nature*, 546(7656), pp. 82–90. Available at: https://doi.org/10.1038/nature22901.

In-na, P. *et al.* (2020a) 'Loofah-based microalgae and cyanobacteria biocomposites for intensifying carbon dioxide capture', *Journal of CO2 Utilization*, 42, p. 101348. Available at: https://doi.org/10.1016/j.jcou.2020.101348.

In-na, P. *et al.* (2020b) 'Loofah-based microalgae and cyanobacteria biocomposites for intensifying carbon dioxide capture', *Journal of CO2 Utilization*, 42, p. 101348. Available at: https://doi.org/10.1016/j.jcou.2020.101348.

Jaap, W.C. (2000) 'Coral reef restoration', *Ecological Engineering*, 15(3–4), pp. 345–364. Available at: https://doi.org/10.1016/S0925-8574(00)00085-9.

Malik, S. *et al.* (2020) 'Robotic extrusion of algae-laden hydrogels for large-scale applications', *Global Challenges*, 4(1), p. 1900064. Available at: https://doi.org/10.1002/gch2.201900064.

Mark Co, J.R. and Culaba, A.B. (2019) '3D Printing: challenges and opportunities of an emerging disruptive technology', in *2019 IEEE 11th International Conference on Humanoid, Nanotechnology, Information Technology, Communication and Control, Environment, and Management (HNICEM)*. Laoag: IEEE, pp. 1–6. Available at: https://doi.org/10.1109/HNICEM48295.2019.9073427.

Mohammadifar, M. *et al.* (2018) 'Power-on-paper: origami-inspired fabrication of 3-D microbial fuel cells', *Renewable Energy*, 118, pp. 695–700. Available at: https://doi.org/10.1016/j.renene.2017.11.059.

Morris, A. (2017) *Dutch designers convert algae into bioplastic for 3D printing*, *Dezeen*. Available at: https://www.dezeen.com/2017/12/04/dutch-designers-eric-klarenbeek-maartje-dros-convert-algae-biopolymer-3d-printing-good-design-bad-world/ (Accessed: 23 October 2020).

Omori, M. and Iwao, K. (2014) *Methods of Farming Sexually Propagated Corals and Outplanting for Coral Reef Rehabilitation; with List of References for Coral Reef Rehabilitation through Active Restoration Measure*. Available at: https://icriforum.org/wp-content/uploads/2020/11/Methods-of-farming-sexually-propagated-corals-and-outplanting-for-coral-reef-rehabilitation-with-list-of-references-for-coral-reef-rehabilitation-through-active-restoration-measure-.pdf

Pang, S., Gao, Y. and Choi, S. (2018) 'Flexible and stretchable biobatteries: monolithic integration of membrane-free microbial fuel cells in a single textile layer', *Advanced Energy Materials*, 8(7), p. 1702261.

Poomathi, N. *et al.* (2020) '3D printing in tissue engineering: a state of the art review of technologies and biomaterials', *Rapid Prototyping Journal*, 26(7), pp. 1313–1334. Available at: https://doi.org/10.1108/RPJ-08-2018-0217.

Ramseur, J. L. (2010) *Deepwater Horizon Oil Spill: The Fate of the Oil*. Washington, D.C.

Shaklova, I. and Cruz, M. (2015) *Living Screen: Robotic Fabrication of Algae Based Gels*. MaterialAbility. Available at: http://materiability.com/portfolio/living-screen/ (Accessed: 27 October 2020).

Stanier, R.Y. *et al.* (1971) 'Purification and properties of unicellular blue-green algae (order Chroococcales)', *Bacteriological reviews*, 35(2), pp. 171–205.

Stefanova, A. *et al.* (2019) 'Approach to biologically made materials and advanced fabrication practices', in M. Asefi and M. Gorgolewski (eds) *International Conference on Emerging Technologies In Architectural Design (ICETAD2019)*. Toronto: Ryerson University, pp. 193–200.

Stefanova, A., Bridgens, B., In-na, P., *et al.* (2020) 'Architectural laboratory practice for the development of clay and ceramic-based photosynthetic biocomposites', *Technology/Architecture + Design*, 4(2), pp. 200–210. Available at: https://doi.org/10.1080/24751448.2020.1804764.

Stefanova, A., Bridgens, B., Armstrong, R., *et al.* (2020) 'Engineering a living building realm: development of protective coatings for photosynthetic ceramic biocomposite materials', in *S.ARCH: The 7th International Conference on Architecture and Built Environment with Architecture AWARDs*. Tokyo: Get It Published Verlag, pp. 362–372.

Stefanova, A. *et al.* (2021) 'Photosynthetic textile biocomposites: using laboratory testing and digital fabrication to develop flexible living building materials', *Science and Engineering of Composite Materials*, 28(1), pp. 223–236. Available at: https://doi.org/10.1515/secm-2021-0023.

Suggett, D.J. *et al.* (2020) 'Coralclip®: a low-cost solution for rapid and targeted out-planting of coral at scale', *Restoration Ecology*, 28(2), pp. 289–296. Available at: https://doi.org/10.1111/rec.13070.

Thimijan, R.W. and Heins, R.D. (1983) 'Photometric, radiometric, and quantum light units of measure: a review of procedures for interconversion', *HortScience*, 18(6), pp. 818–822.

Where to Next?

Chapter 5

DOI: 10.4324/9781003363774-5

Biomaterials are often met with enthusiasm by architects and designers outside the field of bio-design. Bio-design projects oftentimes examine the potentials of biological alternatives. With these potentials, questions arise: why are these solutions not more widely implemented through mass production and commercial products?

The reality is that there isn't a straightforward answer. As the presented research was being conducted, we have seen a greater number of commercial products emerging, such as acoustic panels and packaging by Ecovative and leather by MycoWorks. This gives hope; these early-stage investigations will not remain as novel experiments, simply conducted for the benefit of academics and enthusiasts. However, there are established operational modes of commercial practice, where designers are often compelled to specify products that have been on the market for a prolonged time, which have been previously tested *in vivo* and have demonstrated inert behavior for a prolonged period. This limits the transferability of bio-design output from academia or creative practice into a live scenario. The issue of professional liability is also accompanied by the limitations put in place by project management and procurement structures. Within most projects, delivery timeframes are of the essence; therefore, commercial products that are widely available on the market are chosen in favor of experimental products that are still within the development and testing stage.

There is also the issue of the prohibitive cost of small-scale fabrication and the added costs in custom product manufacture. If manufacture was undertaken by a commercial entity that has an optimized process, then costs would decrease considerably as in the case of engineered brick and other standardized products. Therefore, there are key factors such as public opinion, investment and legislation, to name a few, which are precursors to making biomaterials competitive within a wider context. To make matters more challenging, there are industrial processes designed for the manufacture of conventional static materials, whereas biomaterials come with a different set of manufacturing requirements that necessitate additional outlay costs to transition into a new model of production. The industrialized process of material manufacture is accompanied by contemporary standards for inert materials, which have increasingly longer lifespans and material strength

Where to Next?

that is replaced by dynamic lifecycles as well as reduced durability of biological alternatives. This new materiality goes to the core of the architectural profession, calling for new typologies, revised objectives and localized production, along with a new type of material knowledge that must be cultivated within higher education institutions.

Despite commercial availability presenting a concern, it is important to highlight the importance of early research as it most certainly lays the foundations for scaling and commercialization. In this instance, commercialization refers to making widely available products so as to allow accessibility rather than fitting into capitalist modes of planned obsolescence and standardization. A vast majority of the *in vitro* studies that offer a glimmer of hope and the promise of a sustainable future are often fragile and divorced from the conditions offered by everyday contexts. Therefore, more robust testing with architectural applications in mind becomes key in the transition into mainstream product applications.

The means for integrating academic findings into products is a process that is often undertaken by commercial manufacturers with an economic profit as an incentive. From that perspective, bio-design solutions form a part of a limited market that caters to limited demand. Such efforts are often overshadowed by a fast economy, focused on optimizing production for maximum output. Therefore, these new materials do not yet comfortably fit into the established consumer capitalist model. Unlike products that serve humanity, living organisms demand service and care. Therefore from the point of view of our technology-riddled existence, they are a compromise, rather than a quick fix that can allow us to continue to engage in positive feedback loops (Kimmerer, 2015).

When looking at the potentials of biological material alternatives, it is important to acknowledge that although these new materials and processes can break down small amounts of our waste, absorb some of our pollution and meet a portion of our needs for food, energy and shelter, the reality remains – what can be remedied through these solutions is currently a mere fraction of human needs and impact. Therefore, there is an urgent need to look for ways of managing human behavior, including economic growth in relation to the limitations of our habitat, population growth, consumerism and planned obsolescence.

Furthermore, among architects, questions of scaling and application are often a primary focus, as our discipline traditionally has an affinity toward predetermined outcomes. After all, we envision the world as we would like it to be, by moving inanimate blocks and asserting our vision through the persuasive representation techniques, which become binding document in themselves. However, in bio-design, we take on a facilitator role that subtly initiates processes prone to change and uncertainty. It is a role that requires patience and a willingness to listen to the desires of another species. We can liken this relationship to that of an architect and a client; however, this does not fully consider the elusive bond that begins to form between practitioner and organism over the months and years of tending to a nonhuman entity. As an architect, I often speculated as to potential applications, "final" outcomes, end goals and optimal performance, yet I have learned to hold on more loosely to those aspirations as they inevitably arise from an anthropocentric premise. In place of those desires, I learned to trust in the process that solutions along with new questions will reveal themselves through making with living organisms. This became an integral part of my laboratory practice and allowed me to investigate issues sequentially based on previous findings. This I found out was not unusual in the world of science, where end goals are substituted with small steps toward solving an overarching problem. This is a fundamental difference between architectural and scientific practice. In architecture, if the end goal is not accomplished, the project is often deemed a failure and the process in contemporary commercial contexts is a means to an end. However, by undertaking research through creative practice, the process comes to the forefront and is the means for attaining new knowledge. Therefore, the examples demonstrated in this book offer a sample of holistic bio-design approaches that examine the means of designing for uncertain ecological landscapes. The work is not conclusive, but rather it sets out a cultural shift in bio-design thinking, which I anticipate will continue to evolve and morph into new types of practice, where living organisms are partners in an effort to carve out new ways of living on a planet that no longer offers favorable conditions.

THE CONSCIOUSNESS SHIFT

Through our efforts to regulate our artificial envelopes and meet our physiological needs, we negate climatic conditions and place great strain on the natural environment and resources. Austerity and behavioral changes can only partially reduce that impact; therefore, alternatives to mechanical processes offer a glimmer of hope. While a return to nature, as advocated by some, is unlikely, a closer relationship with nature can aid our journey into renegotiating our place within planetary networks.

Legislation is being put into place to reduce, manage and offset negative environmental effects. However, despite these efforts, an increase of population and greater demand for resources puts pressures on our built environment. Adopted measures that reduce waste and enable reuse of materials are still part of a messy reality of retrieving viable materials, separating the individual building layers and dealing with aging and decay. Through the integration of living metabolisms into the building realm, mechanical processes can be replaced with biological agents that have evolved, adapted, and have the ability to be assimilated back into nature, once they are no longer viable, creating continuous cycles. While the complete replacement of traditional services with biological systems may be challenging, there is a potential for integrating biological systems to partially help address some of the issues associated with the building fabric.

The close relationship between humans and their built environment much resembles that of the skin organ, and as such, buildings can be viewed as a self-sustaining organism that vary their properties to suit a multitude of needs (Mogas-Soldevila and Oxman, 2015). These isolationist efforts are challenged through bio-design where outputs may not be sufficient to address conventional needs, but may interact with the environment, incite other processes to occur or help offset environmentally damaging emissions while bridging the divide created by contemporary modes of urban living. Therefore, there is a need for a philosophical shift to occur to enable us to conceive of materials as living, evolving and self-building blocks, rather than as inanimate matter that possesses qualities to be utilized as tools for human comfort. This new building realm is predicated on fostering a symbiotic relationship with nature, an argument that serves as the basis for the explorations in this book. Rather than drawing hard lines and indiscriminately expelling all other species to sustain and ensure the health of our own, we can begin to look for ways to survive and flourish through collaboration with other organisms. For that to happen, we need to begin to look at the building fabric as an organism, rather than as a machine, and to that end, develop bio-matter that interfaces with nature and our own ecology.

PLURALITY OF PRACTICE

Although bio-design is a field in its infancy, we can be quietly optimistic that it is here to stay. It has only recently emerged, and it is yet to be integrated within the architectural profession. In its short history, the field has been an assemblage of practices and individual philosophies that have been brought under a loosely defined term. This is presenting opportunities in the form of defining new ways of working, outside of an established system, that liberate design practitioners to carve out new investigative paths and to determine their own boundaries of practice. The other side of this argument is the lack of a blueprint that leaves design practitioners without a clear route into bio-design. As previously identified, the challenges lay in the need to venture outside of the design disciplines to accommodate this type of practice. The partnerships that laid the foundations for the work presented in this book can be viewed as living organisms in themselves, which need to be nourished and developed over time. The organic nature of those interactions presented unexpected opportunities for investigation and prevented the creation of a clear plan, prior to initiating the work (Stefanova, 2021).

Creative practice is a particularly appropriate research method, as the making process is a vital component in understanding the nature of working with living organisms (Nelson, 2013). Research through design allows the practitioner to approach problems from multiple angles, which permit the exploration of an emerging phenomenon, through a bespoke set of tools, drawing on the sciences, design, social and political theory, as well as anthropology. Although the bio-design practice presented in this book is a combination of exploratory methods, they do have a commonality in the underlying plurality of architectural thinking. Architectural practice can be viewed as an assemblage of thinking methods that, unlike engineering, are not purely solutions-focused, but rather there is a fascination with the rippling effect of design throughout society. Hence, there is a wider impact and the cultivation of tacit knowledge through working with materials from a creative practitioner's standpoint. This traditional use of both qualitative and quantitative methods in architecture is embraced in the holistic design practice approach advocated in this book, as there was an early recognition that the questions that arose could not be fully addressed through material development alone, as bio-design has a resonant effect within everyday culture that trickles into creative discourse.

CONSTRUCTING NETWORKS: THE RELATIONSHIP BETWEEN SCIENTIFIC AND CREATIVE PRACTICE

Stephen Clark, a philosopher specializing in animal rights, highlights the interconnectedness between humanity and nature in *The Cobra and Other*. He declares that "Human societies everywhere are never wholly human, any more than they are wholly adult, or wholly male, or wholly rational. People, to survive at all, need to understand and work with other creatures, tame and wild" (Clark, 1996). This plurality and diversity can be translated into a plurality of practice, which, in this book, takes on the form of interdisciplinarity, diversity of methods and holistic thinking.

Plurality is laden with pitfalls pertaining to the difficulties in linking natural and social sciences, as identified by Ted Benton. According to Benton, this would require a reexamination of our established views of human and nonhuman relationships.

> There is now widespread agreement that in order to do worthwhile sociological research on the 'material' dimension of environmental issues, the basic conceptual legacy of the sociological traditions has to be radically reworked. In particular, the dualistic oppositions between subject and object, meaning and cause, mind and matter, human and animal, and, above all, culture (or society) and nature have to be rejected and transcended.
>
> (Benton, 1994, p. 28)

A non-dualistic framework is required to avoid the objectification of nature and all of its components. Benton acknowledges that the divide between the social and natural sciences is deeply engrained within our theories and methodologies, and that the challenge of bridging that divides calls for fundamental remodeling of widely accepted concepts. Duality within Western thought places higher value upon abstraction. Notions of increase of capital and market value are perceived as superior considerations, compared to the health and well-being of the ecosystem. This being said, as we enter an era of disturbed ecologies that impact economic interests, we are gradually witnessing a shift in commercial goals to incorporate environmental concerns into business strategies.

The anthropocentric prism that governs our decision-making is also riddled with social barriers pertaining to the microbial world. These constructs of contemporary society have historically categorized fungus, bacteria and plankton as predominantly pathogens and miasmas. This is a notion formulated by early studies into hygiene and perpetuated by the mainstream media. This is in stark contrast to probiotic approaches to design and lifecycle of biological agent that underlies emerging practices. In this context, bio-designers and architects can contribute to this emerging discourse, by offering traditional design skills, alongside a hybridized mode of working that engages the sciences and the humanities.

To illustrate the blurring of divides, Bruno Latour looks at science and its ability to foster networks of relations beyond the confines of the lab for the purposes of disseminating scientific knowledge and artifacts to the rest of society. In his interpretation, Latour rejects dualistic interpretations of natural and social theory, placing an emphasis on the actors that science can engage with beyond the lab as the distinction between inside and outside is replaced by networks that reach across the two realms (Murdoch, 1997). Latour uses the case study of Louis Pasteur efforts to deal with anthrax, a bacterium that was infecting livestock across France in the 1880s. Pasteur set up a laboratory in a farm in rural France to study the bacterium, later transferring the lab back to Paris to develop an effective vaccine. This blurred the divide between science in its designated domain within an institution and formed a relationship that was predicated on an essential collaboration with the outside, belonging to the realm of the social. This collaboration required a level of cooperation, where the lab left its designated place and the farm that hosted Pasteur had to adopt a set of rituals, hygiene, order, quantifying and recording, which came with the physical artifacts. In Latour's words,

> [Pasteur] is master of one technique of farming no farmer knows, microbe farming. This is enough to do what no farmer could ever have done: grow the bacillus in isolation and in such a large quantity that, although invisible, it becomes visible.
>
> (Latour, 1983, p. 146)

Where to Next?

It is important to remember that nature is not present in its raw state within the laboratory; it is translated and manipulated by science before being introduced into the social realm (Murdoch, 1997). Within my research, I am seeking to establish a series of connections, a choreography interlinking the realm of the artificial human, *every day*, and nature as introduced into the artificial through the activities of the laboratory. By working with scientists and learning from the methods they employ within such a setting, I am endeavoring to find viable ways of translating their research into systems capable of functioning outside a controlled environment.

In addition to the scientists and the organisms, it is important to speculate as to the relationship with another actor group, that of the building inhabitant. As we transition into an age of wet living and potentially sentient built environments, humanity is faced with the challenge of reimagining the everyday, redefining the existing value system and bringing critical thinking into making practices that are emerging due to environmental pressures.

Fostering a relationship between the microbial organism and the inhabitants involves development of ways of coexisting and cultivation of empathy, rather than the development of a service module, in relation to a human end user. To avoid the objectification of the living organisms, it is important to regard these living entities as actors capable of affecting the network. This is particularly challenging due to their scale and lack of visible living behavior on a human scale or anthropomorphic features that can give the human actor a sense of recognition.

Although such technologies are still viewed as experimental, oftentimes approached with skepticism and reluctance, they hold an important place within bio-design practice as they propose mending of the links between humanity and nature. The idea that our waste can be useful and that it can become a resource for other species is a fundamental part of bio-design discourse. Furthermore, these contemporary examples look for ways to meet human needs without a negative impact on the environment. Such fledgling research projects encourage a strand of thinking within the scientific and design communities, fueling the imagination of the public and becoming acts of socialization of microbial life within mainstream culture (Figure 5.1).

Figure 5.1 Mycelium tiles with various posttreatments.

BIBLIOGRAPHY

Benton, T. (1994) 'Biology and social theory in the environmental debate', in M.R. Redclift and T. Benton (eds) *Social Theory and the Global Environment*. London: Routledge, p. 271.

Clark, S.R.L. (1996) 'The cobra as other', *Times Literary Supplement*, January, p. 12.

Kimmerer, R.W. (2015) *Braiding Sweetgrass: Indigenous Wisdom, Scientific Knowledge and the Teachings of Plants*. Minneapolis: Milkweed Editions.

Latour, B. (1983) 'Give me a laboratory and I will raise the world', in K. Knorr-Cetina and M. Mulkay (eds) *Science Observed: Perspectives on the Social Study of Science*. London: Sage Publications, pp. 141–170.

Mogas-Soldevila, L. and Oxman, N. (2015) 'Water-based engineering and fabrication: large-scale additive manufacturing of biomaterials', in *MRS Proceedings*. Cambridge: Cambridge University Press.

Murdoch, J. (1997) 'Inhuman/nonhuman/human: actor-network theory and the prospects for a nondualistic and symmetrical perspective on nature and society', *Environment and Planning D: Society and Space*, 15(6), pp. 731–756. Available at: https://doi.org/10.1068/d150731.

Nelson, R. (2013) *Practice as Research in the Arts: Principles, Protocols, Pedagogies, Resistances, Palgrave Macmillan*. London: Palgrave Macmillan.

Stefanova, A. (2021) 'Practices in bio-design: design research through interdisciplinary collaboration', in A. Chakrabarti et al. (eds) *ICoRD 2021: Design for Tomorrow–Volume 3*. 1st edn. Singapore: Springer Nature Singapore Pte Ltd, pp. 41–52. Available at: https://doi.org/10.1007/978-981-16-0084-5-4.

Glossary

Algae – photosynthetic organisms that lack complex organs and tissue and that are commonly found within aquatic or wet environments

Amoeba – a unicellular organism that can change its shape

Autoclave – a laboratory piece of equipment that applies pressure and steam to sterilization

Bio-mineralization – the act of living organisms producing minerals that solidify tissue or encapsulate a substrate

Biofluorescence – fluorescence produced by a living organism

Bioluminescence –light produced by a living organism

Centrifugation – the act of centrifugal force to separate a fluid into various components

Centrifuge – an apparatus that uses centrifugal force to separate a fluid into various components

Chitosan – a polysaccharide made by treating the shells of crustaceans with an alkaline chemical, such as sodium hydroxide, commonly used to separate algae from water through a process of flocculation

Chlorophyll fluorescence – light reemitted by plants, algae and bacteria used as an indicator of photosynthetic energy conversion.

Compound microscope – a microscope that uses multiple lenses

(Microscope) cover slip – A very tiny piece of glass that is placed over a specimen on a slide to flatten the sample and aid microscope imaging

Cyanobacteria – photosynthetic bacteria that are some of the oldest organisms on the planet

Dinoflagellate – a type of single-celled phytoplankton species, some of which bio-luminesce when agitated

Extrusion 3D printing – Printing using pressure rather than heat, typically using a viscous substance

Flocculation – chemical reaction that results in the adhesion of particles into clusters and therefore resulting in sedimentation

Fluorescence – light emitted by an object that has absorbed it

FTIR (Fourier transform infrared spectroscopy) – a type of spectroscopy that captures the infrared spectrum of substances in various states, including gas, liquid and solid

Fungus – a group of eukaryotic organisms that include yeasts, molds and mushrooms

HEPA filter (high-efficiency particulate absorbing filter) – a certain standard of air filters

Hydrogel – a hydrophilic polymer that does not dissolve in water that can, in certain instances, be used as a matrix for cultivation of microorganisms

Hyphae – a strand-like, branching fungal structure

Incubation – the period of growth after inoculation

Inoculation – introduction of a microorganism into a growth media or a substrate

Kappa-carrageenan – extracted from red seaweed; it is used to galette a solution

Laminar flow hood – an enclosed cabinet that pumps in air through a HEPA filter to create a clean zone for aseptic work

Matrix (material matrix) – the substance within which a something lives or resides

Microporous tape – a breathable tape that allows air exchange but prevents the entry of pathogens

Monoculture – the practice of growing a single species at a time

Mycelium – a branching fungal structure made up of hyphae, which is typically found in soils and within the fruiting body of the fungus

Myco-welding – the binding or fusing of mycelium composites done by growing them while in contact with one another, resulting in the mycelium hyphae growing into a single mesh

Outplanting – the process of planting an organism following growth within an artificial environment

Parafilm – a translucent thermoplastic used to seal containers to prevent contamination

Pasteurization – the process of using moderate heat levels to kill pathogens

pH (potential of hydrogen) – a measure of the acidity or alkalinity of an aqueous solution

Photosynthesis – the conversion of light energy into chemical energy

Photosystem II – a protein supercomplex present in the membrane of photosynthetic plants that is responsible for initial reaction of photosynthesis

Phytoplankton – photosynthetic microorganisms that include algae and cyanobacteria

I-PAM – Imaging Pulse Amplitude Modulation Fluorometry

Scanning electron microscope (SEM) – a microscope that produces an image by scanning a surface with an electron beam

Spectroscopy – the study of how electromagnetic spectra are affected by the interaction between electromagnetic radiation and matter

Stereo microscope – a low-magnification microscope that allows larger areas to be viewed often within a petri dish

Tissue culture – growing parts of plant or animal tissue in an artificial setting, separate from the organism from which it originated

Index

μmol 110
absorb 120
absorbance 98
absorbed 14, 129
absorbing 33, 98, 130
abundant 56, 65, 67
ACADIA 16, 72
ACDP 29, 44
acetic 69, 70, 71
Acetobacter 66, 67, 71, 73, 74, 75, 77
acid 41, 60, 69, 70, 71
acidic 70
acidified 61
acidity 130
acoustic 56, 119
adhesion 102, 129
adhesive 56, 94
Aerobacter 66
Aeronautics 27
aesthetics 67
Agar x, 38, 51, 52, 53, 54, 55, 60, 69
Agarikon 21
aggregate 14, 107
agitated x, 67, 74, 129
agitating 68
agitation 52, 66
agricultural 6, 56, 64, 75
Agrobacterium 66, 75
airflow 81
airtight 111
AJMR 73
alcohol 9, 30, 31, 32, 34, 71, 82
algae ix, 11, 12, 13, 17, 24, 30, 31, 34, 39, 43, 49, 53, 57, 58, 59, 60, 61, 62, 63, 64, 65, 72, 73, 76, 77, 78, 88, 90, 97, 98, 103, 106, 107, 110, 111, 112, 115, 116, 129, 130
algaebiopolymer 115

algaeladen 44
Algaerium 12, 20
algal 11, 19, 20, 45, 73
alkaline 70, 129
alkalinity 130
aloe 98
aluminum 53
ammonium 60, 63
amoeba 129
amount 11, 13, 17, 36, 37, 39, 57, 58, 60, 68, 97, 113, 120
amplitude 43, 44, 46, 103, 130
analysis iv, 9, 37, 39, 41, 43, 71, 74, 75, 111
analytical 45
analyzed 37, 41
analyzing 41
animal xiv, 5, 6, 9, 14, 16, 17, 124, 131
anthropocene ii, 17, 21, 46, 86, 97, 115
anthropocentric 121, 125
anthropogenic 3
antibiotic 5, 9
antifungals 56
antioxidants 73
apparatus 43, 129
application xii, xiv, 5, 7, 9, 10, 11, 12, 18, 19, 20, 24, 36, 45, 49, 61, 66, 71, 72, 74, 75, 97, 98, 103, 111, 115, 120, 121
approach ix, xvi, 4, 14, 15, 17, 21, 23, 64, 70, 75, 88, 96, 116, 121, 123, 124, 125
aquatic 65, 129
aqueous 130
ARCC 21, 46
Archaea 17
architect ii, 3, 12, 13, 97, 119, 121, 125

Index   133

architectural i, 4, 17, 21, 23, 45, 46, 77, 114, 116, 120, 121, 123

architecture ii, iv, xii, 8, 15, 16, 17, 20, 21, 23, 24, 25, 26, 46, 50, 77, 114, 116, 121, 124

area iv, ix, 28, 30, 31, 33, 39, 41, 43, 52, 53, 54, 55, 57, 65, 66, 85, 89, 98, 107, 131

artificial 6, 11, 58, 61, 70, 115, 121, 126, 130, 131

artificially 6

artist ii, 7, 98

artistic 11

arts 16, 20, 27, 127

aseptic ix, 28, 31, 35, 45, 52, 54, 55, 76, 130

asking 33

aspen 49, 57

ATCC 50, 72

Atlantis 3, 16

atmosphere 11

atmospheric 9

AURO 98, 101, 102, 103, 104, 105, 106, 110, 111, 112

author vi, xiv, 26, 27, 44, 82, 83, 84, 85, 86, 87, 88

autoclave 28, 29, 30, 51, 53, 55, 129

autoclaved 51, 54, 59, 60, 64, 69

autoclaving 53, 56, 57

autonomous 15, 77

autopoiesis 4, 5

autotrophic 57, 58, 61, 76

autotrophically 57

average 12, 63, 65, 93

avoid 41, 124, 126

Azotobacter 66

Bacillus 67, 125

bacterial x, 5, 9, 10, 11, 16, 18, 20, 30, 34, 39, 43, 64, 65, 66, 67, 68, 69, 70, 71, 72, 73, 74, 75, 76, 77, 78, 98

bacterium ix, 3, 9, 10, 15, 16, 20, 31, 32, 49, 64, 66, 74, 76, 125, 129

base 65, 67, 70, 91, 110

baseline 110

BCCM 50, 72

beaker 53, 110

beam 131

beehives 19

benchmark 90, 93

benchmarking 90

benefit xiv, 5, 6, 20, 25, 42, 49, 58, 61, 63, 64, 89, 119

binder 98, 103, 105, 106, 110, 111, 112

binding 121, 130

biobatteries 76, 116

biocellulose 73

biochemical 74

biochemistry 73, 77

biocompatible 77

biocomposite 21, 46, 72, 74, 77, 78, 97, 115, 116

BioCouture 10

biodegradable 7, 13

biodesign iv, xii, 17, 25, 120, 122

biodiesel 57, 72, 73, 75, 76, 78

biodiversity 36, 50

bioengineered xiv

bioengineering xiv, 72, 73, 74, 75, 76

biofilm 17, 43, 44, 73, 88

biofluorescence 129

biofuel 3, 64

biohybrid 17, 115

bioinspired iv

biological ii, iv, xii, 5, 10, 13, 15, 28, 29, 36, 39, 40, 59, 67, 75, 103, 119, 120, 122, 125

biologically 21, 116
biologists 3
biology xii, xiv, 9, 16, 17, 18, 19, 21, 44, 45, 127
bioluminescence 11, 129
bioluminescent 10, 20, 59
biomass 5, 11, 12, 13, 57, 58, 72, 74, 76, 78
Biomaterialia 115
biomaterials 14, 98, 116, 119, 127
biomedical 66, 71, 114
biomimetic 19, 20, 97
biomimicry iv, 8, 20, 21
biomineralization 10, 11, 19, 64, 97, 115
biomineralized 13, 14
biomolecules 72
bioplastic 115
bioprinting 114
bioprocessing 72
bioproducts 16
bioreaction 74
bioreactor 12, 66, 68
bioremediation 12, 13, 15, 61, 63, 73, 78
bioremediators 36
bioresource 16, 45, 72, 77, 78
biortech 45, 72, 77, 78
biosafety 29, 45
bioscience 74
biotech 74
biotechnical 23
biotechnological 78
biotechnology ii, xii, xiv, xvi, 18, 25, 45, 72, 73, 74, 77, 78
bisque 109, 110
bleaching 70, 86, 114
blocks 3, 7, 121, 122
blue 17, 41, 59, 72, 111
bluegreen 110, 116
blueprint 123

boiling 57, 69
bovine 10
breathable 130
breeding 88
brewing 69
brick 7, 21, 119
brittle 66, 70, 104
broader xii, 23
broth 51, 61, 69, 73
Bunsen 28, 29, 31, 32
burner 28, 29, 31, 32
byproducts 5

CaCl 60
calcite 5
calcium 10, 51
canister 101, 102
carbohydrate 19, 42, 57, 75
carbon 5, 7, 11, 13, 14, 17, 36, 37, 57, 58, 61, 69, 74, 75, 78, 115
carbonate 10
carboys 30, 59
carrageenan 98, 101, 102, 103, 104, 106, 110, 111, 112, 130
cartridges 110
carve 121, 123
cast 81, 82, 83, 86
casting x, 6, 16, 49, 50, 56, 72, 81, 82, 83, 85, 86
catalysts 7
CCAP 60, 72
cell ix, 5, 6, 10, 16, 19, 20, 21, 37, 39, 41, 42, 43, 44, 45, 57, 58, 64, 65, 73, 74, 76, 98, 103, 104, 107, 108, 112, 115, 116
celled 97, 129
cellulose x, 5, 10, 11, 18, 20, 30, 43, 49, 55, 65, 66, 67, 68, 69, 70, 71, 72, 73, 74, 75, 76, 77, 78, 98
centrifugal 65, 110, 129

centrifugation 65, 129
centrifuge 30, 75, 129
centrifuged 110
ceramic xi, 43, 44, 46, 77, 94, 97, 106, 109, 111, 112, 113, 114, 116
characterisation 18
characteristic 18, 19, 74, 76
characterization 16, 72, 99
characterized 41
characterizing 42
chemical 3, 10, 18, 27, 28, 36, 39, 42, 56, 61, 63, 64, 65, 66, 70, 72, 74, 75, 76, 129, 130
chemistry 20, 72, 76
chemists 3
chitosan 101, 129
Chlorella 57, 59, 63, 74, 75, 76, 78, 98, 107
chlorophyll 43, 44, 46, 103, 104, 105, 107, 129
chromium 63
cider 69
CITA 25
citrate 60, 69
citric 60, 71
clay xi, 24, 40, 46, 77, 85, 88, 95, 98, 101, 102, 104, 106, 107, 108, 110, 111, 114, 116
clean 31, 46, 130
cleaning 27, 28, 30
cleanliness 25
climate xiv, 83, 96, 97
climatic 121
coagulants 65
coat 29, 34
coated 112
coatings 66, 101, 116
coexistence 16
coexisting 126

coffee 7, 56, 69
collaborate xvi
collaborating 27
collaboration ix, 25, 46, 122, 125, 127
collaborative 25, 26, 27
collaborator xvi
colonization 16, 38, 49, 55, 56, 81
colonize 55, 85, 97
colonized 35, 49, 54, 55, 82
colonizing 49
colony 38, 88
coloration 70
compatibility 36, 37, 81, 98, 101, 107
compatible 6, 52, 63, 85, 110
complex 3, 9, 15, 20, 129
complexity xiii, 5, 23, 106
component i, 13, 14, 23, 97, 107, 112, 114, 123, 124, 129
composite xi, 8, 16, 18, 20, 21, 44, 77, 97, 98, 104, 105, 106, 114, 116, 130
composition 9, 17, 37, 45, 61, 63, 64, 66, 67, 69, 101
compostable 7, 66
compound 7, 41, 129
compression 82
compressive 7, 37, 56, 83
concentrated 59, 60
concentration 53, 65, 66, 70, 73
concrete 10, 11, 18, 81
condensation 54
conditions ix, 5, 11, 12, 14, 31, 35, 36, 37, 38, 52, 58, 64, 68, 69, 70, 77, 78, 81, 95, 102, 111, 120, 121
conducive 26
consistency 53, 59, 64, 85, 95, 101, 102, 104, 107, 110

constraints 4, 85
construct 33, 125
constructed 15, 33
constructing xi, 124
construction iv, 5, 6, 11, 14, 15, 17, 18, 35, 45, 77
consumables 27, 29, 30, 31
containment 29, 44
contaminants 52, 57, 78
contaminate 53
contamination ix, 25, 27, 31, 34, 35, 43, 49, 51, 57, 58, 59, 69, 81, 101, 130
control ix, 28, 29, 30, 38, 39, 40, 67, 103, 108, 111, 114, 115
converted 57
copper 63
coral xi, xvi, 85, 86, 88, 89, 93, 94, 95, 114, 115, 116
Coralclip 94, 116
CoralPlug xvi, 93, 94, 95, 96
cost 13, 27, 33, 49, 50, 58, 61, 63, 93, 116, 119
cotton 41, 66, 98, 99, 103, 104, 105
coverslip 41
COVID 24
coworking 26
craftsman 23
crosslinked 71
crustaceans 101, 129
CSIRO 59, 73
cultivate 51, 55, 59, 63, 69, 78
cultivated 6, 12, 13, 26, 40, 49, 53, 57, 58, 59, 61, 111, 120
cultivating 5, 8, 67
cultivation x, 6, 13, 35, 36, 51, 53, 57, 58, 59, 61, 64, 67, 68, 69, 72, 74, 75, 76, 78, 93, 107, 123, 126, 130

cultivator 9, 21
culture ix, x, 5, 6, 12, 16, 17, 27, 30, 35, 36, 37, 38, 39, 41, 42, 45, 49, 50, 51, 52, 54, 55, 58, 59, 60, 61, 62, 64, 65, 67, 68, 69, 72, 73, 74, 75, 76, 77, 78, 98, 107, 108, 110, 124, 127, 131
cultured 6, 20
curvature 82
curve 33, 34
curved 82, 83
CuSO 60
cyanobacteria x, 9, 12, 41, 42, 44, 57, 61, 72, 74, 97, 115, 129, 130
cyanobacterial 42
cycle 15, 37, 59, 69, 109, 110, 111, 122

data vi, 4, 37, 38, 43, 103, 111, 114
daylight 110
days 28, 34, 38, 43, 49, 55, 63, 68, 70, 104
dead 5, 15, 39, 43, 55
decompose 14
decomposed 7
decomposers 7
decomposition 43
decrease 67, 68, 119
decreased 57, 69, 95
Deepwater 116
degree 37, 51, 71, 81
deionized 28, 29, 42, 51, 53, 60, 69
deletion 67
demolition 11
deposition 93, 98
deposits 50
derived 6, 63, 74

Index 137

design i, ii, iv, vi, ix, x, xii, xv, xvi,
3, 4, 5, 7, 8, 11, 12, 13, 15, 17,
18, 19, 20, 21, 22, 23, 24, 25,
26, 27, 28, 29, 30, 31, 32, 33,
34, 35, 36, 37, 38, 39, 40, 41,
42, 43, 44, 45, 46, 49, 50, 51,
56, 59, 66, 67, 72, 77, 80, 81,
82, 83, 84, 85, 86, 87, 88, 89,
90, 91, 92, 93, 94, 95, 96, 97,
98, 99, 100, 101, 102, 103,
104, 105, 106, 107, 108, 109,
110, 111, 112, 113, 114, 115,
116, 119, 121, 123, 124, 125,
126, 127
designated 28, 32, 81, 125
designed vi, 10, 49, 68, 90, 119
designer i, iv, v, 3, 4, 7, 8, 9, 10,
11, 14, 23, 24, 25, 26, 28, 45,
96, 98, 115, 119, 125
designing ix, xiv, 17, 19, 36, 37,
81, 121
deviation 37, 99
device xii, 65
dextrose 57
diatoms 97
diffusion 17
digest 7
digital ii, 7, 21, 41, 77, 85, 98,
114, 116
dihydrate 51
diluted 42, 59, 60, 64, 74, 95
dilution 37, 40, 41, 64, 110
dinoflagellate 59, 129
diodes 43
dioxide 5, 11, 36, 37, 57, 58, 61,
74, 115
disinfect 34, 82
disinfected 27
disodium 69
displaying 5

disposable 29, 31, 34
disposal ix, 27, 28, 29, 35
dispose 27, 31
dissolve 130
dissolved 74, 107, 110
distilled 107
distortion 41, 95, 102
diversity xv, 11, 18, 21, 74, 124
drying x, 65, 66, 70, 71, 72, 82,
83, 104, 105
DSMZ 50, 73
durability 14, 120
duration 111
dwelling 45
dyeing 70
dyes 63, 66, 70
dying 15
dynamic 15, 76, 120

Earth 7, 21, 71
eCAADe 114
ecological ii, 7, 10, 17, 115, 121
EcoLogicStudio 12, 13, 18
ecology 8, 11, 16, 19, 20, 76, 86,
88, 97, 116, 123, 125
ecosystem 88, 125
Ecovative 49, 50, 87, 119
EDTANa 60
efficiency 33, 57, 65, 78, 130
efficient 12, 49, 65
effort xiv, 25, 120, 121, 122, 125
elasticity 70
electrical 12, 13, 14, 16, 54
electricity 11
electrochemical 75
electrode 75
electrolytic 65
electromagnetic 131
electron 42, 44, 131
electronic vi, 72

138   Index

elegance xiii
elegans 6, 16
element 3, 13, 15, 23, 81, 83
emission 11, 17, 42, 122
emitted 129
emitting 43
encapsulate 7, 97, 129
encapsulation 97
energy 10, 13, 14, 19, 58, 64, 65, 72, 73, 75, 76, 115, 116, 120, 129, 130
engineered 11, 19, 61, 115, 119
engineering xii, 18, 19, 21, 23, 72, 74, 75, 76, 77, 78, 97, 115, 116, 123, 127
engineers 3, 45, 74
enriched 53, 57, 69
environment ii, xiv, xvi, 4, 5, 6, 10, 11, 12, 14, 15, 16, 19, 21, 23, 25, 26, 27, 30, 32, 36, 38, 39, 45, 46, 51, 53, 58, 64, 65, 69, 82, 83, 85, 86, 88, 97, 98, 103, 106, 107, 112, 115, 116, 122, 126, 127, 129, 130
environmental i, ii, xiv, 11, 13, 21, 27, 36, 45, 59, 63, 71, 72, 74, 78, 97, 122, 124, 125, 126, 127
environmentally 122
epidorphs 30
equipment 13, 25, 27, 28, 29, 30, 31, 33, 35, 41, 42, 51, 57, 110, 129
ethanol 30, 34, 69, 71
ethical xiv, xv, 4
ethics ix, xiv, xv, 97
eukaryote 17
eukaryotic 57, 129
EVAC 18
evaporation 71, 103, 112, 113, 114

evolution 9, 16, 19
evolutionary 6, 16
evolve 121
evolved 10, 11, 122
evolving 9, 122
example xii, 4, 5, 7, 8, 9, 11, 12, 13, 14, 15, 16, 25, 27, 29, 33, 34, 36, 38, 39, 41, 53, 56, 58, 59, 60, 61, 63, 68, 69, 70, 82, 95, 97, 114, 121, 126
exhibit 13, 23, 38
exhibited 6, 103, 111
existence 9, 120
exoskeletons 101
experiment ix, x, xii, xiv, 4, 6, 9, 24, 30, 37, 38, 39, 45, 49, 61, 64, 65, 76, 80, 81, 82, 83, 84, 85, 86, 87, 88, 89, 90, 91, 92, 93, 94, 95, 96, 97, 98, 99, 100, 101, 102, 103, 104, 105, 106, 107, 108, 109, 110, 111, 112, 113, 114, 115, 116, 119
experimental i, ii, ix, 9, 11, 12, 13, 15, 19, 23, 24, 36, 37, 38, 44, 71, 81, 98, 119, 126
experimenting 67
exposure 112
external 66
extinctions 6
extract 13, 17, 51, 53, 55, 57, 69, 72
extracted 51, 101, 106, 130
extraction 64
extremophiles 36
extrude 24, 85
extruded 102
extrusion x, 7, 85, 96, 98, 102, 110, 115, 129
eyepieces 41

Index 139

fabric ix, 12, 13, 14, 15, 66, 97,
106, 122, 123
fabricate 25, 95
fabricating 6, 8, 97
fabrication ii, 4, 6, 7, 8, 11, 15,
19, 21, 49, 56, 77, 78, 85, 87,
92, 95, 98, 103, 114, 115, 116,
119, 127
facade 20
façade 5, 12, 15, 76
factor 14, 35, 36, 38, 64, 66, 67,
68, 111, 112, 119
falcon 30, 54, 110
fashion iv, xii, 66
fermentation x, 4, 7, 9, 66, 67, 68,
70, 73, 75, 76, 77, 78
fermented 9
ferric 60
fibers 77
fibrils 75
fiction iv, xii, 10
filament 102
film 53, 66, 77, 81, 82
filter 12, 28, 30, 33, 34, 51, 52, 55,
57, 65, 82, 130
filtered 64, 65
fixation 11, 57
flame 31
flammable 29
flasks 30
flexibility 31, 70
flexible 12, 21, 76, 77, 82, 116
floatation 65
flocculants 65
flocculation 65, 76, 129
florescence 43
flourish 11, 107, 122
flowhood 46
fluid 45, 129
fluorescence 37, 43, 44, 46, 103,
104, 105, 107, 129

fluorescent 110
fluorometry 44, 46, 103, 130
foil 53, 54
folds 112
footprint 7
force 65, 110, 129
formulating xv
formwork 18, 81, 83
fouling 88, 90
Fourier 42, 45, 129
fraction 120
fragile 15, 104, 120
frame 14, 83, 84, 104
framework iv, 17, 124
freeze 65, 71, 74
freezer 29
freshwater 74
fridge 29
fructose 69, 76
FTIR 42, 45, 129
fuel 10, 11, 19, 58, 64, 76, 115, 116
function 5, 7, 9, 10, 12, 14, 15, 21,
25, 29, 36, 63, 97, 106, 111
functional 20, 31, 115
fundamental 4, 5, 8, 19, 36, 37,
53, 83, 114, 121, 124, 126
funding 25, 27
fungal x, 6, 16, 17, 18, 20, 36, 39,
41, 42, 43, 44, 45, 49, 50, 51,
55, 66, 130
fungi 7, 8, 9, 17, 19, 21, 31, 35, 43,
45, 59, 65, 75, 97
fungus ix, x, 6, 7, 15, 18, 30, 49,
51, 73, 114, 125, 129, 130
future ii, 3, 10, 11, 15, 17, 21, 28,
38, 51, 75, 120

Ganoderma 56, 77, 81
gasses 42
gels 101, 116
gene 3

genera 66, 67
general 7, 29, 31, 72
generate 4, 11, 110
generated 11, 16, 65, 109
genetic 11, 16, 17
genetically 3, 35
genomic 16
genus 43
glass 30, 51, 53, 54, 55, 107, 129
glassware 30, 59
glimmer 120, 122
glove 29, 31, 34
glovebox 45
glow 3
gluconacetobacter 73, 75
glucose 69, 74, 75
glutamate 77
glycerol 69, 70, 71
goggles 29
grain x, 53, 55, 56
gravitational 65
gravity 82
green 15, 17, 19, 20, 59, 60, 69, 72
greenhouse 63, 75
grow 6, 7, 12, 26, 43, 49, 53, 55,
    56, 64, 66, 81, 89, 125
growing i, iv, ix, 4, 5, 6, 7, 8, 10,
    11, 12, 13, 15, 18, 21, 23, 37,
    46, 50, 52, 53, 55, 57, 66, 70,
    76, 98, 130, 131
grown 6, 7, 16, 49, 51, 52, 55, 57,
    58, 66, 70, 71, 82, 106, 110
growth 6, 7, 12, 20, 30, 50, 52, 53,
    55, 58, 61, 63, 65, 66, 67, 68,
    70, 74, 77, 81, 88, 94, 95, 107,
    111, 120, 130
gypsum 51, 55, 57

habitable i, 3, 13
habitat ii, 11, 14, 85, 97, 120
Haematococcus 59

hardwoods 56
harvest 12, 78
harvested 7, 64
harvesting x, 64, 65, 70, 72, 75, 76
HBBE ii
hemicellulose 55
hemp 66
HEPA 33, 34, 130
hessian 98, 100, 103
heterotrophic 57, 58, 76, 78
hollow 81, 107, 112
holobionts 9
homogeneity 51, 56, 76, 85, 110
honey 9
honeybees 19
hood 28, 31, 33, 34, 35, 45, 46, 130
hospitable 9, 56, 91, 112
host 6
hosted 125
humidifier 30, 114
humidity 7, 37, 59, 81, 82
hybrid 4, 49, 58, 125
hydrated 55
hydrocolloid 73, 101, 105
hydrogel xi, 44, 71, 97, 103,
    115, 130
hydrogen 65, 70, 130
hydrophilic 130
hydroponics 15
hydroxide 41, 69, 70, 129
hygiene 4, 125
hygrometer 30
hygroscopic 105
Hylozoic 15
hyphae 7, 41, 42, 130
hypothesis 37

ICMP 50, 74
ICSA 115
ICXN 21
illuminate 64, 111

Index   141

imaged ix, x, 39, 41, 42, 43, 44, 45, 103, 111, 112, 129, 130
immobilising 5, 72
impermeable 112
inanimate 3, 5, 23, 121, 122
incentive 120
incubate x, 5, 6, 7, 30, 34, 37, 38, 39, 43, 52, 55, 59, 68, 69, 81, 82, 87, 103, 104, 110, 111, 130
incubator 30, 59, 66
indicator 129
indigenous 127
infecting 75, 125
infrared 42, 45, 129
ingredient 60, 63
inhabitant 23, 113, 126
inject 51, 52, 55
inkjet 12
inks 114
inner 19
inoculate 31, 38, 39, 55, 77, 110, 130
inoculum 78
inorganic 3, 97, 106, 115
instrument 10, 27, 41, 44
interact xii, 23, 37, 38, 112, 122, 123, 131
interdisciplinarity i, xvi, 25, 46, 124, 127
interior 31, 66, 98, 114
interlinking 126
interlocking 113
invented vi
ions 65
isopropyl 30, 34

JAMB 73
JMEST 76
JNUS 71
JOBAB 16
juvenile 88, 94, 95

kappa 101, 102, 104, 106, 110, 111, 112, 130
kelp 14
kiln 96, 109
Komagataeibacter 66, 67, 72, 75, 77
kombucha 67, 68, 69, 71, 73, 74

laboratory i, 5, 21, 25, 26, 27, 31, 36, 39, 44, 45, 46, 49, 50, 51, 54, 61, 63, 64, 76, 77, 93, 103, 116, 121, 125, 126, 127, 129
labs ix, 13, 19, 23, 27, 33
lactate 76
lactic 41, 69
lacto 41
lactophenol 41
laminar 28, 31, 33, 34, 35, 45, 46, 130
landscape 6, 8, 10, 86, 97, 121
larvae 88, 89, 90, 93
larval 114
laser 83, 87
LasX 41, 42, 99
layer 13, 14, 52, 70, 76, 82, 97, 108, 109, 110, 113, 116, 122
leather 6, 7, 10, 17, 119
leave 4, 54, 123
lens 3, 14, 26, 43
lenses 30, 41, 44, 129
level 3, 7, 23, 25, 29, 34, 37, 38, 39, 43, 44, 45, 52, 58, 59, 66, 67, 90, 95, 104, 107, 112, 114, 125, 130
liability 119
life 3, 7, 9, 10, 11, 14, 15, 16, 19, 20, 21, 23, 45, 64, 74, 119, 127, 130
lifecycle 23, 120, 125

142   Index

lifespans 14, 66, 119
light 6, 7, 10, 12, 17, 30, 37,
    39, 41, 43, 51, 53, 57, 58,
    59, 66, 110, 111, 112, 116,
    129, 130
lighting 10, 64
lightweight 12
lignin 55
lignocellulosic 18
linen 103
lining 20
lipid 42, 57, 76
liquid x, 12, 15, 35, 36, 41, 42, 49,
    51, 52, 54, 55, 58, 60, 61, 64,
    65, 67, 78, 110, 129
livestock 125
living i, iv, xvi, 3, 5, 6, 7, 8, 12,
    13, 14, 15, 16, 18, 19, 21, 23,
    25, 28, 31, 38, 39, 44, 45, 49, 50,
    54, 60, 64, 77, 81, 82, 83, 85,
    87, 97, 103, 104, 106, 107, 114,
    115, 116, 120, 121, 122, 123,
    126, 129
load 18
lockdown 24
loofah 74, 106, 115
lucidum 56, 57, 77, 81
luminesce 129
Lutum 95, 102, 110

macromolecules 75
magnesium 57
magnetic 30, 51, 52, 53
magnification 39, 41, 42, 99, 131
maintenance 13, 15, 28, 30, 75
malt 51, 53, 55, 57
manufacture 12, 33, 88, 91,
    114, 119
manufacturers 120
manufacturing 5, 7, 20, 73,
    119, 127

marin 12, 26
marine xvi, 3, 20, 86, 98
masonry 10, 13
mass 5, 6, 15, 51, 52, 53, 55, 74,
    81, 119
matrix 24, 37, 56, 85, 97, 98, 101,
    102, 103, 104, 106, 110, 111,
    112, 130
matt 19, 54, 115
matter ix, 3, 4, 5, 12, 15, 16, 19,
    23, 36, 39, 55, 71, 115, 119,
    122, 123, 124, 131
mechanical vi, 4, 6, 10, 13, 14,
    15, 16, 18, 23, 65, 76, 102,
    106, 122
mechanically 12
mechanisms 25
mechanistic 3, 15
medium x, 3, 15, 28, 30, 39, 40,
    41, 49, 51, 52, 54, 59, 60, 61,
    63, 64, 66, 67, 69, 70, 72, 73,
    74, 110, 125, 130
membrane 65, 75, 76, 77, 78,
    116, 130
mesh 41, 130
metabolic 5, 7, 9, 10, 11, 63,
    66, 97
metabolism 5, 15, 122
metabolize 36
metagenomic 74
metal 30, 36, 60, 63, 94
meter 30, 69, 107, 108
MgSO 60
miasmas 15, 125
MicoWorks 7
microalgae x, 12, 19, 36, 40,
    41, 44, 57, 58, 59, 61, 63,
    64, 72, 73, 74, 75, 76, 78,
    114, 115
microalgal 45, 75, 107
microbe 19, 21, 125

Index   143

microbial 3, 7, 9, 10, 11, 15, 16, 19, 23, 39, 45, 64, 73, 74, 76, 115, 116, 125, 126, 127
microbiological 44, 45, 46
microbiology 17, 46, 73, 78
microbiome 9, 21
microfiltration 76
microflotation 73
microns 33
microorganism i, 5, 8, 9, 12, 29, 39, 49, 50, 66, 69, 72, 73, 74, 76, 97, 115, 130
micropipettors 31, 45
microporous 51, 52, 130
microscale 10, 99
microscope 10, 30, 37, 39, 41, 42, 43, 45, 99, 129, 131
microscopic 45
microscopy 39, 42, 44, 45
microstructural 16
microwaving 71
migration 24, 44, 103
mineralization 10, 129
mineralized 16, 72
minerals 129
minimal 33, 58, 65, 106, 107
Minitab 38
Mistry 77
mitochondrial ii
Mixotrophic 57, 58, 76
MLCB 41
MnCl 60
modulation 43, 44, 46, 103, 130
module 87, 88, 126
moisture 14, 30, 58, 65, 82, 83, 95, 98, 106, 107, 114
molasses 69, 73
mold 49, 81, 82, 83, 87, 129
molecular 17, 71, 75
monoculture 50, 51, 130

monokaryon 20
monolithic 56, 76, 116
monotub 28, 31, 32
multicellular 5, 9, 18, 57
multidisciplinary 26
multifunctional 18
multispecies 11, 19, 69
mushroom 4, 7, 8, 9, 17, 20, 21, 30, 49, 52, 55, 57, 77, 91, 94, 96, 129
mycelial 19, 20, 41, 54
myceliation 84
mycelium x, xvi, 5, 7, 8, 14, 16, 17, 18, 21, 41, 49, 51, 52, 53, 55, 56, 69, 71, 77, 81, 82, 83, 84, 85, 86, 87, 88, 98, 114, 127, 130
Myco 7, 8, 81, 88, 130
mycological 18
mycology 8, 9, 18, 19, 73, 75
mycorrhizal 36, 50
MycoWorks 20, 119

NaNO 60
nanocellulose 72, 74
nanotechnology 5, 115
NASA 27
nature i, 3, 8, 14, 16, 19, 20, 21, 25, 46, 50, 73, 97, 104, 112, 115, 122, 123, 124, 126, 127
Naumann, A. 43, 45
NCIM 50, 76
needle 30, 51, 52
network xi, 4, 15, 16, 25, 50, 122, 124, 125, 126, 127
neutral 107
nitrate 61
nitrile 31, 34
nitrogen 63, 64, 69
nondistractive 43

nondualistic 127
nonhuman 23, 97, 121, 124, 127
nonliving 106
nonporous 66, 70, 81
nonsacrificial 75
nonscientific 31
nonspecialist 25
nozzle 96, 101, 102, 103, 110
nutrient x, 6, 12, 15, 31, 34, 37,
   39, 50, 51, 53, 55, 58, 59, 61,
   63, 64, 66, 68, 69, 74, 75, 78,
   98, 107, 112

observation 37
observational 37, 38
observe 38, 54
observed 39, 52, 127
observing 43
obsolescence 14, 120
occupancy 28
ocean 86, 114
octopus 10
optimal 28, 39, 64, 67, 68,
   101, 121
optimistic 123
optimization 72, 73, 74, 75, 77, 78
optimize 95
optimized 49, 58, 61, 119
optimizing 74, 120
organ 6, 45, 122, 129
organic 7, 12, 26, 57, 58, 65,
   97, 123
organism x, xi, 3, 5, 6, 7, 8, 9, 12,
   14, 15, 16, 18, 19, 23, 25, 27,
   28, 34, 35, 36, 37, 38, 39, 43,
   44, 50, 53, 54, 57, 59, 81, 85,
   91, 97, 106, 120, 121, 122, 123,
   126, 129, 130, 131
organismal 5, 10, 53
origami 19, 116

orthographic 107
ostreatus 57, 77
outcome 8, 10, 23, 36, 38, 49, 121
outplanted 94, 95
outplanting 93, 115, 116, 130
output 12, 15, 26, 44, 111, 119,
   120, 122
oven 7, 71, 83
oxygen 5, 9, 11, 12, 63, 66, 67, 68,
   74, 81
oxygenation x, 67, 68
ozone 57, 65

paint 13, 17, 98, 101, 102, 104,
   106, 110, 111
paintbrush 110
pandemic 24
panels 14, 17, 119
paper 3, 19, 38, 66, 81, 83, 87,
   97, 115
parafilm 30, 52, 53, 54, 130
parameters 7
parametric 8, 20
particle 31, 33, 49, 52, 56, 129
partitions 13, 98
partnerships i, 5, 11, 50, 123
paste 85
Pasteur, L. 125
pasteurization 130
pasteurize 53
pasteurized 57
patented 3
patents vi
pathogenic 9
pathogens 3, 15, 125, 130
patterns 16, 104
pedagogies 127
pellicle 67, 69, 70
peptone 69
perforated 81, 82

Index   145

perforating 81

peroxide 70, 77

petri x, 4, 30, 34, 39, 53, 54, 55, 59, 61, 69, 103, 131

petrochemical 13

pharmacology 76

phosphate 63, 69

phosphorous 63, 64

phosphorus 63

photobiology 17

photobioreactor 12, 72

photochemistry 17

photoheterotrophic 57, 58

photometric 116

photosynthesis 12, 57, 130

photosynthetic x, xi, xvi, 5, 21, 39, 43, 44, 46, 57, 59, 77, 91, 97, 104, 106, 114, 116, 129, 130

photosystem 43, 130

phototropic 12

phycology 76

physicochemical 74

physics 21, 78

physiological 9, 121

physiology 76

phytoplankton 11, 13, 20, 59, 61, 65, 76, 129, 130

plankton 15, 20, 73, 125

planned 120

plant 5, 6, 12, 15, 16, 44, 45, 50, 57, 63, 64, 65, 66, 72, 73, 74, 98, 106, 111, 127, 129, 130, 131

plaster 81

plastic 30, 31, 53, 57, 70, 81, 82, 83, 86, 110, 111, 114

plasticizer 71, 77

plasticizing 70, 71

plating 31, 45, 54, 61, 76

Pleurotus 57, 77

podium 39

pollutants 11, 63

pollution 11, 12, 86, 114, 120

polyester 98, 99, 103

polymer 7, 19, 65, 71, 74, 75, 76, 77, 98, 106, 130

polymerizing 74

polysaccharide 19, 20, 129

polyvinyl 71

porosity 78, 107, 109

porous 70, 81

posttreatment x, 66, 67, 70, 127

potassium 41

powder 53, 73, 110

powered 65, 77

predator 95

predatory 89

predetermined 121

prefabricated 81, 104

presets 95

pressure 6, 29, 30, 33, 34, 51, 76, 85, 95, 101, 102, 110, 122, 126, 129

pressurization 102

pretreated 56

probe 45, 99, 107

probiotic 10, 15, 19, 125

procurement 119

prokaryote 16

prokaryotic 57

propagation 56, 57, 75, 114

protein 42, 130

protocol xiv, 4, 26, 28, 29, 35, 37, 38, 44, 45, 70, 98, 106, 127

prototype 8, 11, 12, 16, 38, 84, 86, 95

prototyping 116

protozoa 60, 72

pulse 43, 44, 46, 103, 130

pump 30, 81, 130
purification 20, 70, 71, 116
purifying 70
purity 59, 66
Puspitasari, P. 77
puts 122
pyramid 44, 107, 109

qualitative 37, 124
quantifiable 71
quantifying 125
quantitative 37, 43, 124
quantity 58, 126

radiation 131
range i, iv, xii, 7, 8, 10, 11, 51, 61, 66, 67, 68, 81, 104
ratio 49, 57, 60, 64, 77, 110
reabsorption 71
reactor 72
readings 38, 107
reagents 37
recipe 51, 60, 69
recycling 78
reef 88, 94, 95, 114, 115, 116
refeeding 69
refrigerator 28, 52
regeneration 15, 19, 115
regions 81
rehabilitate 115
rehabilitation 115, 116
rehydrated 41
rehydration 70, 103, 104
remediate 36
remediating 3, 8, 17
remediation 11, 63, 78
remediators 7
remedied 120
renegotiating 122
renewable 19, 66, 72, 73, 115

repair 5, 10
repeatability 38
repeatable 37, 51
replicates 37
representation 23, 121
reproductive 10
requiring 12, 25, 67
research i, ii, iv, xiv, xvi, 4, 6, 8, 17, 18, 19, 21, 24, 25, 26, 27, 28, 36, 38, 45, 46, 50, 56, 63, 66, 73, 75, 76, 78, 106, 114, 119, 120, 121, 123, 124, 126, 127
researcher iv, 6, 12, 25, 37, 50, 88
researching 8
reshape i, 10
residency 26
resilience 10, 97, 112
resilient 49, 88
resistance 33, 34, 127
resonant 124
resource 7, 12, 26, 59, 72, 74, 122, 126
responsive 10, 49, 50
restoration xi, 85, 88, 95, 96, 97, 114, 115, 116
restoring 88
restrictions 6, 25
result xiv, 11, 37, 38, 39, 49, 58, 61, 63, 64, 67, 68, 70, 88, 97, 99, 101, 103, 104, 109, 111, 114, 129
resulted 27, 28, 83, 95
resulting 11, 57, 95, 129, 130
retardation 113
reuse 122
rewetting 70
Rhizobium 66, 71
RIBA 20
rice 55, 56, 69, 77
risk xiv, 27, 28, 41, 51, 59

Index   147

rituals 125
robotic 18, 77, 115, 116
RStudio 38

safe 28
safely 28, 31
safety ix, 27, 28, 29, 30, 36, 46
salt 9, 63
sample ix, 27, 28, 37, 38, 39, 41, 43, 44, 49, 50, 51, 53, 54, 71, 103, 107, 108, 109, 111, 112, 121, 129
sand 10
sanitized 52
saturated 112
saturation 46
sawdust 56
scale 4, 7, 13, 18, 25, 29, 63, 64, 75, 98, 115, 116, 119, 126, 127
scaling 8, 120, 121
scalpel 52
scanning 42, 44, 131
science ii, ix, xii, xvi, 4, 8, 10, 11, 16, 17, 18, 19, 20, 21, 23, 25, 26, 27, 29, 33, 42, 71, 74, 75, 76, 77, 78, 97, 114, 116, 121, 123, 124, 125, 126, 127
scientific xi, 4, 9, 10, 19, 25, 26, 27, 28, 38, 49, 50, 67, 88, 93, 98, 114, 121, 124, 125, 126, 127
scientists i, 3, 5, 7, 10, 11, 27, 28, 96, 126
scoby 67, 69, 71
screen 116
screening 78
seaweed 20, 98, 106, 130
sediment 64, 88, 90
sedimentation 64, 65, 75, 129
seeds 55
selected 33, 36, 52, 63, 88, 98, 107

selfgenerating 15
senescence 50
sensationalized 3
sense xvi, 3, 15, 126
sensing 16
sensors 10
sensory 10
sentient 126
sequence 37, 86
sequenced 50, 67
sequester 5, 11, 12, 13, 58, 106
series iv, ix, xii, xiii, xiv, xv, xvi, 21, 43, 71, 78, 110, 126
serve 14, 26, 38, 67, 107, 120, 122
service 13, 120, 122, 126
sets 107, 109, 111, 121
setting ix, 4, 9, 15, 23, 24, 25, 26, 28, 31, 32, 33, 36, 37, 38, 39, 50, 56, 57, 59, 63, 64, 66, 81, 83, 93, 95, 104, 126, 131
settlement 93
setup 4, 12, 24, 25, 33, 41, 51, 61, 62, 108, 110, 113, 114
sexually 115, 116
shadowing ix, 25, 26
shaking 55, 66, 68, 77
shallow 53
sharps 28, 29
sheet 66, 67, 69, 70, 71, 81, 82
shell 7, 13, 97, 108, 129
shrinkage 104
SICI 72
sieve 55
signage 30
significance 23, 37, 38, 61, 63, 68, 71, 111, 112
simulation 20, 38
simulator 41
sink 29
site 12, 13, 15, 90, 93

size 30, 31, 33, 39, 49, 56, 65, 78, 102
skill 4, 26, 45, 97, 125
skin 12, 20, 55, 73, 122
slant x, 53, 54
slanted 54
Slicer 95, 110
slide 30, 41, 129
slip 30, 129
sludge 77
slurry 12, 58, 65, 102, 110, 111
soaked 44
soap 70
socialization 127
socialize 15
society ix, xiv, xv, 9, 15, 17, 86, 123, 124, 125, 127
sociological 124
sodium 61, 69, 70, 77, 129
software 3, 38, 41, 42, 99, 110
soil 11, 15, 19, 130
solid 42, 60, 69, 77, 97, 129
solidify 54, 129
solution 4, 6, 8, 11, 12, 13, 14, 36, 52, 53, 54, 59, 60, 63, 64, 65, 67, 70, 75, 96, 103, 116, 119, 120, 121, 123, 130
solve xiv, 3, 4, 36
source x, xii, 8, 9, 11, 25, 28, 31, 35, 36, 49, 50, 55, 56, 57, 58, 61, 64, 66, 69, 86, 114
sourced 49, 51, 59, 64, 65, 69, 95, 107
sourcing x, 27, 29, 49, 51, 59, 66
space ix, 5, 12, 23, 25, 26, 27, 28, 29, 30, 31, 35, 36, 52, 58, 59, 69, 81, 97, 108, 127
span 10
spark 26
spatulas 30

spawn x, 49, 51, 54, 55, 57
special 95, 107
specialist 13, 25, 26, 67
species ix, x, 5, 6, 7, 8, 9, 11, 12, 14, 15, 18, 23, 28, 30, 35, 36, 39, 41, 49, 50, 51, 52, 54, 55, 56, 57, 58, 59, 61, 63, 66, 67, 75, 82, 86, 121, 122, 126, 129, 130
specimen 50, 51, 52, 129
specimens 27, 50, 51
spectra 42, 131
spectroscopy x, 37, 42, 45, 129, 131
spectrum 23, 42, 50, 67, 129
speculate 8, 126
speculated 121
speculation xii, 6, 14
speculative ii, 8, 11, 13
speed 21, 55, 77, 81, 95
spikes 89
spill 86, 116
spirit xvi
spirulina 59
spore 7, 31, 42
spray 31, 103
sprayed 82, 111
SPSS 38
stable 23, 103
stack 54, 75
stacking 81
stage 26, 35, 55, 58, 70, 88, 98, 112, 119
stagnation 50
stains 41
standard 23, 27, 33, 37, 49, 69, 93, 95, 99, 119, 120, 130
standpoint 71, 124
star 89, 95
starfish 90, 96

start 14

static x, 5, 15, 67, 68, 74, 81, 119

statistical 37, 38

StDev 99, 100

steam 129

steel 14

stereo 39, 41, 131

stereotyped 9

sterile 28, 30, 31, 32, 34, 45, 53, 54, 55, 57, 59, 129

sticker 51

stimulate 43

stimulators 73

stirrer 30, 51, 53

stock 59, 60

stoneware 43, 95, 107

storage vi, 27, 28, 81, 103, 104

stored 28, 41, 51

strain x, 38, 39, 43, 59, 66, 67, 68, 75, 122

strand 97, 126, 130

strategy i, 74, 113, 125

stream 6

strength 7, 37, 56, 61, 109, 110, 119

Streptococcus 17

structural 5, 7, 8, 10, 13, 14, 16, 17, 18, 20, 21, 26, 42, 43, 72, 77, 81, 83, 84, 88, 91, 95, 97, 98, 102, 106, 107, 113, 114, 119, 130

student x, xvi, 26, 50, 59, 82, 83, 84, 85, 86, 87, 88

studied 6, 25, 37, 44, 66

studio xvi, 23, 27, 32, 82, 83, 84, 85, 86, 87, 88, 95, 110

study 4, 5, 7, 8, 10, 12, 19, 36, 37, 38, 44, 45, 53, 58, 64, 67, 68, 70, 77, 93, 97, 98, 101, 104, 106, 107, 109, 111, 113, 114, 120, 125, 127, 131

styrofoam 14

subculture 51, 53, 54, 55, 59, 60

subfield 7

subject 10, 72, 124

submerging 70

substance 28, 42, 54, 129, 130

substitute 10, 12, 13, 63

substrate x, 7, 18, 28, 35, 37, 39, 40, 42, 43, 49, 55, 56, 57, 64, 74, 77, 81, 82, 88, 93, 97, 98, 106, 107, 111, 114, 129, 130

sucrose 69, 73

sugar 6, 57, 58, 69

sulfate 51, 57

supplement 13, 57, 77, 127

supplemented 28, 50

suppliers 31, 33, 49, 59

supply 6, 49, 51, 56, 62, 66, 68, 81

support xii, xvi, 9, 15, 17, 23, 37, 82

surface x, 34, 42, 44, 53, 54, 55, 64, 65, 66, 67, 70, 71, 73, 74, 82, 88, 97, 102, 104, 107, 112, 113, 131

survey 29, 57, 59

survive 50, 122, 124

suspension x, 41, 42, 57, 58, 65, 111

sustain 36, 53, 64, 103, 106, 122

sustainability 7, 15, 64, 81

sustainable 8, 11, 17, 18, 19, 20, 23, 64, 66, 73, 75, 106, 120

sustained 13

sustaining 5, 23, 88, 97, 122

symbiosis 5

symbiotic 9, 16, 50, 122

symmetrical 127

synthesis 72, 74

synthesized 75

synthetic xii, 20, 61, 64, 69, 97

syringe 30, 49, 51, 110
system iv, vi, x, 4, 12, 13, 15, 18, 19,
    20, 23, 29, 45, 51, 58, 65, 73, 77,
    93, 98, 106, 114, 122, 123, 126

tablets 61
tank 58, 64, 65, 93
tape 30, 51, 52, 130
target 59
targeted 116
task 27, 36
teaching ii, 45, 127
team 26, 88, 93
technical 114
technique i, xiv, 6, 9, 10, 31,
    37, 38, 43, 44, 45, 54, 55, 76,
    121, 125
technocratic 23
technological xii
technology xii, xiv, 6, 10, 12, 15,
    16, 17, 18, 21, 25, 45, 46, 64,
    72, 73, 74, 75, 76, 77, 78, 115,
    116, 120, 126
Teflon 71
temperature x, 7, 30, 36, 37, 43,
    52, 54, 59, 66, 68, 71, 73, 81,
    107, 109, 110, 111, 112, 114
tensile 37, 70, 98, 113
tension 82
tent 30, 81
terrestrial 57, 65, 86
territory xii, 8, 20, 21
test 4, 16, 21, 27, 34, 37, 38, 39,
    51, 77, 85, 90, 93, 96, 98, 101,
    102, 103, 107, 108, 110, 111,
    116, 119, 120
textile xi, 10, 11, 12, 19, 20, 21,
    63, 66, 73, 76, 77, 82, 83, 85,
    86, 97, 98, 99, 102, 103, 104,
    106, 116

texture 56, 66, 71, 81, 85, 98
textured 95
theory ii, xii, 17, 123, 124,
    125, 127
therapeutic 19
therapy 73
thermal 13
thermoplastic 130
thickness 67, 70, 106, 107, 108,
    109, 112, 114
threshold 38
tiles 127
timber 6, 7, 14, 19, 33, 49, 56, 57,
    66, 81, 83, 84
time ii, x, 3, 5, 13, 24, 26, 27, 28,
    37, 38, 41, 43, 44, 49, 51, 52,
    55, 57, 58, 63, 65, 66, 68, 71,
    73, 81, 82, 83, 92, 93, 95, 104,
    105, 119, 123, 127, 130
timeframes 119
timer 30
timescale 16
tissue xii, 6, 17, 20, 45, 116,
    129, 131
tool 4, 25, 30, 34, 37, 39, 60, 74,
    114, 122, 123
toxic 56, 63
toxicity 108
toxins 3
transfer 31, 37, 45, 51, 52, 53,
    54, 55, 58, 60, 61, 62, 74,
    119, 125
translucent 66, 112, 130
transparency 77
transparent 12
transportation 103, 104
treatment 9, 20, 57, 70, 71, 76, 77,
    78, 110
tree 6, 12, 57
trends 88

Index   151

trial 101
triplicate 38, 103, 111
tube 30, 53, 54, 59, 110
tubing 30, 59, 70, 81
tubs 30
tunability 74
tunable 16
turbulent 16, 72
tweezers 30, 52

UKRI 27
uncontrolled 5, 102, 103
underperformance 11
unicellular 9, 116, 129
unified iv
uniform 71
unit 83, 113, 116
untreated 57, 81
upcycle 14
uploads 72, 116
upon 25, 45, 51, 52, 54, 71, 95, 102, 109, 124
uptake 26
upward 31, 89, 93
urea 64
urine 64, 74, 77
UTEX 59, 78

vaccine 125
vacuum 81, 82
value xiv, 37, 38, 124, 126
vapor 114
variable 4, 37, 38, 39, 93, 95, 105, 112
variation 38, 49, 63, 90, 107
variety 7, 49, 97
vegetal 72
vegetation 15

vegetative 7
vein 12
velocity 33
vernacular 14
vessel 30, 81, 106, 109, 110
viability 38, 68
viable 17, 20, 36, 65, 93, 96, 114, 122, 126
view i, ix, 3, 4, 9, 10, 14, 15, 16, 17, 21, 23, 28, 39, 41, 42, 54, 97, 120, 122, 123, 124, 126, 131
vinegar 69
viscosity 102
viscous 56, 85, 129
visibility 31
visible 38, 39, 43, 52, 54, 103, 112, 126
visibly 43
vision 121
visual 16, 43, 44, 103, 111
visualized 45, 76
visually 4, 111
vital 9, 13, 36, 39, 114, 123
vitamin 75
vitro 4, 36, 37, 120
vivo 4, 64, 113, 119
voices iv
volume i, 31, 45, 46, 49, 56, 58, 81, 127
volumetric 33, 34, 45
VormVrij 95, 101
vulgaris 57, 58, 59, 61, 63, 64, 74, 75, 76, 98, 102, 103, 107, 110
vulnerable 58

Walz 43
wash 55, 56, 70, 77, 112
wastage 6

waste ix, 6, 7, 11, 14, 15, 27, 29, 30, 36, 56, 63, 64, 76, 78, 120, 122, 126
wastewater 11, 57, 58, 63, 64, 69, 73, 76, 77, 78
water 11, 12, 13, 15, 28, 29, 33, 41, 42, 51, 53, 54, 55, 57, 58, 60, 63, 64, 65, 66, 69, 70, 71, 75, 77, 78, 93, 95, 107, 108, 110, 111, 112, 114, 127, 129, 130
wavelengths 43
ways i, xiii, 4, 31, 38, 70, 83, 89, 93, 98, 120, 121, 122, 123, 126
WCST 18
wearable 64, 77
weight 16, 98

welfare xiv
WFCC 59, 78
workshop 24, 32, 56

xylinum 66, 71, 72, 73, 74, 75
xylinus 73, 75, 77

yarn 13
years 3, 9, 14, 121
yeast 51, 52, 53, 55, 57, 69, 129
yield 13, 58, 68, 69, 73
yielding 51
yogurt 9

zinc 63
ZnSO 60
zone 31, 32, 34, 35, 54, 130

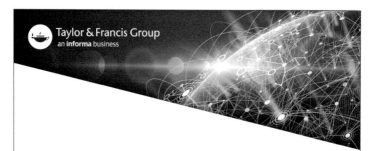